THE
QUIRKS & QUARKS
QUESTION BOOK

THE
QUIRKS & QUARKS
QUESTION BOOK

101 ANSWERS TO
LISTENERS' QUESTIONS

INTRODUCED BY
BOB MCDONALD

CBC radioONE

M&S

National Library of Canada Cataloguing in Publication Data

Main entry under title:

The Quirks & quarks question book:
101 answers to listeners' questions

ISBN 0-7710-5448-3

1. Science – Miscellanea. I. McDonald, Bob, 1951-

Q162.Q57 2002   500   C2001-903833-X

We acknowledge the financial support of the Government of Canada through the Book Publishing Industry Development Program for our publishing activities. We further acknowledge the support of the Canada Council for the Arts and the Ontario Arts Council for our publishing program.

Typeset in Palatino by M&S, Toronto
Printed and bound in Canada

McClelland & Stewart Ltd.
The Canadian Publishers
481 University Avenue
Toronto, Ontario
M5G 2E9
www.mcclelland.com

1   2   3   4   5   06   05   04   03   02

CONTENTS

▲

INTRODUCTION BY BOB McDONALD

▲

Imagine a world without good questions. It would be a very small and boring place indeed, without curiosity, devoid of interest, and, worst of all, a world without the *Quirks & Quarks Question Show*. The hundred and one brain-bafflers and puzzlers in this book are just a sample of the countless queries that have arrived in our office over the last decade. Rooting through what used to be the mailbag, now increasingly the inbox on our e-mail system, we look for the quirkiest questions. You know, the kind that puts a smile on your face, or a wrinkle on your brow, even the kind a child might ask: "Mom, why doesn't Dad wake himself up with his own snoring?" You don't find the answer to a question like that in a book (except this one, of course). You need to find an expert, and that's where the fun begins for us. We comb the universities and scientific institutions of this country, looking for the right person, a search that often involves dozens of phone calls, in a version of telephone tag. Faced by what on the surface appears to be a simple question, it's quite common for a scientist to say, "Gee, I hadn't thought of that. I have no idea what the answer is, but I know someone who might." And on the quest goes, often skipping from coast to coast until the mystery is finally solved. Then we put it on the program, either as part of our award-winning *Question Show* specials, or on the regular "Question of the Week" segment.

Good science is asking good questions. Astronomers point telescopes at the sky to answer "How high is up?" Biologists peer down microscopes, asking, "How do single cells become complete human beings?" Physicists smash sub-atomic particles together in huge accelerators, seeking the answer to "How small can anything actually get?" We have questioned the nature of our world and discovered that it is definitely not what it seems.

Without scientific inquiry, or, for that matter, just basic human curiosity, we would perceive nature only through our five senses, where we can see no farther than the horizon, no higher than the clouds, no smaller than a grain of sand. From that narrow perspective, the world appears mostly flat and unmoving, with a giant dome, half blue, half black, continuously rolling above us. Plants simply appear out of the ground, and babies, well, they just happen. For millions of years, our ancestors lived in this small bubble of perception. Descartes, the seventeenth-century philosopher, advised us to question everything and believe only what can be proven systematically, setting the foundation for modern scientific thought. Now, thanks to centuries of asking good questions, we have expanded our bubble of perception to include an extremely large universe, and it is still expanding! We have looked into the leaf of a plant and seen cells dividing, over and over again, the very process of life. We have looked into the cell of a human, and discovered the DNA sequences that regulate the Book of Life. When we looked for the smallest particle inside an atom, we found that everything is made of quarks. All these discoveries were the result of asking fundamental questions about the nature of life and the universe.

Of course, that doesn't mean it's all over. We may be pretty smart, but ignorance still abounds. Remember, 90 per cent of the mass in the universe is still missing! But that's good. It keeps scientists busy.

For more than twenty-five years, *Quirks & Quarks* has been following scientists' efforts to answer questions. It has been an amazing experience, riding along the cutting edge of scientific knowledge, sharing the excitement of new discoveries, learning more about how things actually work. During my ten years as host of the program, an amazing list of new discoveries in every area of science has unfolded: Dolly, the cloned sheep, was born; the entire human genome was mapped; new fossils illuminated our past; new journeys to space illuminated, perhaps, our future. We learned that the universe is not only expanding, it's speeding up and probably won't stop; planets have been found orbiting other stars; a small space vehicle landed on Mars; the Mir space station crashed to Earth. Here in Canada, a neutrino observatory opened underground in Sudbury; an ancient iceman was found frozen in the Rockies; four Canadians have flown in space and a new Canadarm was added to the International Space Station.

Throughout this odyssey, we have asked the Canadian scientific community to do a little extra work, by applying their problem-solving skills to answering questions sent in by *Quirks & Quarks* listeners. Every Saturday, on CBC Radio One, half-a-million Canadians from coast to coast to coast (and thousands more listening around the world on short-wave and on the Internet) tune in to hear about the latest scientific discoveries. Many of those listeners wait eagerly for the last segment in the show, when we feature the "Question of the Week." All of the questions in this book have appeared on the program during the past ten seasons, and the scientists and researchers who answer them represent every province and every scientific discipline. We even went back to each expert for updates, in case the answers needed new details.

So, if you have ever wondered how sperm knows which way to go, where we actually live in our galaxy, or whether insects ever sleep, you'll find all that and more here in the *Question Book*. If you have a question that isn't answered in this book, send it in: it may appear on a future show – or even in the next book. There are never enough good questions, and the best ones have come from you.

Thanks, and enjoy!

1

Rivers, Rocks, and Raptors:

GEOGRAPHY, GEOLOGY, AND ARCHAEOLOGY

SANTA CLOCK AT THE NORTH POLE

▲

What time is it at the North Pole?

Dr. Rob Douglas, Physicist with the Time and
Frequency Standards Group at the National
Research Council in Ottawa:

The answer depends on which way you are facing. And the
time on your watch will depend on which route you took to
get there. If you came north from Edmonton, you'd likely be
on Mountain Time; but if you came from Russia, then that's
the time you'd use. The twenty-four time zones, including the
one that's split into two dates by the International Date Line,
all, in principle, meet at the North Pole. As you turn around,
your time will change; but that would keep you very busy
with your watch.

Since this could cause some confusion, there is a better way of telling time at the North Pole. That's to use Coordinated Universal Time (UTC). The definition of UTC is based on what used to be called Greenwich Mean Time. That was based on noon being calculated as the average time when the sun crossed the Prime Meridian at Greenwich, England.

Today, modern technology lets us calculate UTC very accurately. There are two parts to calculating it. One is the longitude of where you are, and the other is the exact time.

As far as longitude goes, the old-fashioned "line in the dirt" at the Greenwich Observatory is no longer good enough for some purposes. You need a world average that takes into account the drift of the continents. And, of course, they are all moving in relation to one another, so there is an agreed-on international coordinate system that lets you find out within centimetres exactly where on the Earth you are. So from the international system we know exactly where the Prime Meridian is.

When it comes to time, we now use atomic clocks. These are so sensitive that, rather than just adding leap days, they add leap seconds to correct for the rotation of the Earth. There is an international reference system that's based on the average of all the world's laboratory atomic clocks. This will tell you exactly what time it is. If you are going to the North Pole, just use Coordinated Universal Time.

MAKE MINE HALF-AND-HALF

▲

I got out my old National Geographic *atlas and my world globe the other day. Looking at the images of the continents and how they have moved over the last two hundred million years, I took some tissue paper and traced the outlines of the present continents in the atlas and placed them on the globe. When I fit them all together into one landmass, I found that this super-continent seemed to occupy exactly half the globe. Are there any theories that suggest that the Earth was at one time covered exactly half with water and half with land?*

Dr. Stephen Kissin, Professor and Chair of the
Geology Department at Lakehead University:

At the present time, the Earth is about one-quarter land and three-quarters water. And various theories suggest that, in fact, there would have been somewhat less land when the land was all in one mass than there is at present. But when you look at a modern map, it does look like it's half-and-half. The problem is the use of a flat projection in the atlas, which distorts the size of the continents, particularly towards the poles.

Since the world is a globe and a map is flat, some of the sections have to be stretched to make the whole thing fit on a piece of paper. The type of map we commonly use is called a Mercator projection. On this map, most of the distortion happens closest to the poles, making Canada, Russia, and Baffin Island look much larger than they really are.

There was a point in geological history when all the land on the planet was assembled into one super-continent. It was between 350 and 260 million years ago, and the continent was called Pangea. However, it wouldn't have been large enough to make the planet half land and half water. What we know about plate tectonics suggests the amount of material making up the continents increases over time. Moreover, there were probably more submerged continental shelves at the time of Pangea than there are now. So, if anything, there was less visible land during the existence of Pangea than at present.

Pangea may not have been the only super-continent on Earth. At one time we thought it was, but a Canadian geologist, the late J. Tuzo Wilson of the University of Toronto, hypothesized that it had happened more than once. His evidence suggested that, in fact, super-continents broke apart and joined together in a cyclic fashion, which came to be called the Wilson cycle.

FLIPPING THE POLES

▲

If the Earth's magnetic field is protecting us from cosmic rays, what happens when the Earth's magnetic field flips? Is there a time when the Earth doesn't have any magnetic field, and what would this mean for life on the planet?

Dr. Larry Newitt, Scientist with the Geomagnetism Program at Natural Resources Canada:

The magnetic field of the Earth surrounds the planet and has poles, like a bar magnet. We call them the North and South Magnetic Poles, and they determine the direction in which a compass needle will point. This magnetic field protects us from the deadly cosmic rays that come hurtling towards us from space.

The Earth's magnetic field certainly has reversed many hundreds of times in the past. When it reverses, the North Magnetic Pole becomes the South Magnetic Pole and vice versa. This last occurred about 780,000 years ago. During a reversal, the Earth's magnetic field goes fairly close to zero, dropping to about 10 per cent of its present value. It's actually a fairly rapid event, on the geological time scale, taking about four thousand years. So we'd be without the shielding properties of the magnetic field only for that period.

The whole process of field reversal is fairly complicated, and there are a couple of competing theories about what really happens during a reversal. One is that the magnetic fields sort of fade out, and during the middle of the reversal, the Earth's magnetic field no longer looks like a bar magnet, but becomes much more complex. Then it grows up again, like a bar magnet, but in the opposite direction. The other competing theory is that it stays like a bar magnet, but gets weaker, and the magnetic poles actually migrate from north to south and from south to north, respectively.

What causes the Earth's magnetic field to flip is not entirely understood. We know the outer core of the Earth is composed primarily of fluid iron. This region is a good conductor of electricity, and the place where the magnetic field is generated. There has to be some subtle change in the motion of the fluid in the outer core to cause the fields to reverse, but what this change is exactly is not clear.

Back in the 1960s, someone put forward the idea that, during a reversal, the protective force of the Earth's magnetic field would disappear and we would be exposed to much higher levels of cosmic rays, which would have all kinds of nasty ramifications for life on Earth. But there doesn't seem to be any evidence that this has been the case. There are a couple of problems with this idea. First of all, even now the Earth's magnetic field doesn't completely protect us from cosmic rays. It sort of shunts the rays around and causes them to penetrate in the polar regions. So people in the Arctic get exposed to more cosmic rays than people at the Equator do, and this doesn't seem to be doing them any harm. During a reversal, the entire Earth would be exposed to the same level of cosmic rays, but you wouldn't expect anything particularly nasty to happen.

The other potential problem is the solar wind, the charged particles that stream out from the sun. This is deflected by the Earth's magnetic field and, again, comes down in the polar regions. But the solar wind only penetrates into the upper atmosphere, and doesn't come down to ground level. We have to remember that, even if the Earth's magnetic field were very weak and not providing much protection, the atmosphere would still cover us. The other thing worth remembering is that, during the last reversal, our humanoid ancestors were around on the Earth, and they seem to have gone through it without any difficulties.

THE AGING EARTH

▲

What will the Earth look like millions of years from now?

Dr. Paul Robinson, Professor in the Department of Marine
Geology and Earth Sciences at Dalhousie University:

We can only speculate on what Earth will look like in the
future, but we can use the past as a guide. In the past, the continents and oceans were arranged in very different patterns
from those of today. At least twice in geological history the
continents came together to form a huge super-continent surrounded by a super-ocean. The first super-continent formed
about 750 million years ago (in late Precambrian time) and the
second about 290 million years ago. The more recent super-
continent was called Pangea.

We can predict some future changes in the distribution of
continents and oceans from modern geological processes. For
example, the Atlantic Ocean is growing larger at a rate of two
to four centimetres per year, as a result of spreading on the
Mid-Atlantic Ridge. This translates into about twenty to forty
kilometres per million years. That doesn't seem like a very
rapid rate of change, but a million years is only a short period
of geological time. In 100 million years (which is still not very
long in geological time) the Atlantic Ocean could grow several
thousand kilometres in width.

In the Pacific Ocean, the sea floor is growing along the East
Pacific Rise, but some of the ocean crust is being consumed in

subduction zones. So part of the Pacific Ocean is actually decreasing in size and North America is slowly moving closer to Asia. The Pacific Ocean is very large, so it will be many millions of years before the two continents ever come into contact. However, it is quite possible that, in the future, the continents of today will form another super-continent surrounded by another super-ocean. This would cause many changes in ocean circulation and global climates.

In the Pacific Ocean there is also one small piece of California that is moving northward. This piece lies on the west side of the San Andreas Fault and is moving at an average rate of a few centimetres per year. Give it another hundred million years and what is now western California may be nicely lodged up into Alaska.

The surface of the Earth is extremely changeable. It may seem stable, but geological time involves tens, hundreds, and even thousands of millions of years. With that much time, even slow processes can cause great changes in the distribution of continents and oceans, in the heights of mountains, in the nature of climates, and in the plants and animals that live on the Earth.

SUCCULENT SAUROPODS
OR DUNKIN' DINOS
▲

What did dinosaurs taste like?

Dr. Hans-Dieter Sues, Vice-President, Collections and
Research at the Royal Ontario Museum, and Professor
of Zoology at the University of Toronto:

Sadly, there is not really a scientific answer to this, so the best
we can do is speculate, based on what we know about the diet
of dinosaurs and what various animals taste like today. People
generally assume that if something is not a mammal, it must
taste like chicken. Whenever you ask people what lizard tastes
like, or snake, or spider, that is the universal response. But this
is only partly true. I have eaten many odd things over the
years, travelling in some of the developing countries around
the world while doing fieldwork. Iguana, for example, does
taste a little like chicken. But many reptiles don't. They have
either very individual flavours or they taste like very salty
chicken or something of this sort. One of my more memorable
dining experiences was in southern China, where I was
offered something that was proudly declared to be land eel
and quite tasty. But as I dissected this land eel to get chopstick-
sized pieces, I discovered that there were little vertebrae in it
that were not fish vertebrae. They were snake vertebrae, and
I figured out that it was a cobra. It tasted quite good, but it did
not taste like chicken.

I suspect that most dinosaurs didn't taste all that great.
One reason would be that the plant-eating dinosaurs, at least
during much of the evolution of dinosaurs, subsisted on
things such as ferns, conifers, and other plants that, if you feed
them to modern animals, make the meat taste really funny.
One of my colleagues once said that one of those large dinosaurs would have tasted like really tough chicken that had

been basted for a very long time in a pine needle marinade. It would not be a delightful culinary experience. Of the meat-eating dinosaurs, the fish-eaters would likely be quite tasty, and the young ones might have been tastier, or at least more tender. Of course, trying to get a young *Tyrannosaurus* out of its nest or away from its mother might be a bit of a challenge. You'd probably end up as dinner yourself.

Unfortunately, apart from these guesses, we will probably never know. Only mineralized pieces of soft tissue from dinosaurs have been preserved, that we know of, and even if dinosaur meat were preserved somehow, it might not taste as it did 65 million years ago. In fact, the only large prehistoric animals whose meat has been preserved are the mammoths and woolly rhinoceroses that have been found buried in snow and ice. One of my old professors once was invited to a dinner by fellow paleontologists in Moscow and they served him Siberian mammoth, which was about fifteen thousand years old. The meat was very well preserved, in terms of its superficial appearance, but he said it was the most awful thing he had ever put in his mouth. Evidently, refrigerating mammoth for that long doesn't improve its flavour.

A SALTY TALE

▲

Why is the ocean salty? And why aren't lakes or rivers salty? What about those few salt lakes?

Dr. Steve Calvert, Professor Emeritus in the Department
of Earth and Ocean Sciences at the University of
British Columbia:

To begin with, you need to understand that the saltiness of
the ocean is not just from sodium chloride – table salt. It is a
mixture of many chemical constituents, or salts in general,
dissolved in seawater. So the question is where are these
chemicals coming from. There are two answers.

The first involves the water cycle on the Earth's surface. In
general, rainwater that falls on land is naturally slightly
acidic. This dilute acid solution, which we call rain, chemi-
cally weathers the rock on the Earth's surface, and eats away
at it, basically dissolving it. As a result, when the rainwater
runs off the land, it is carrying chemicals from the rocks, such
as mineral salts.

The runoff is a very dilute solution of salts, so dilute we
can't taste it. This solution then enters streams and rivers and
eventually flows into the ocean. Water is then evaporated
from the ocean into the atmosphere, leaving the salts behind.
Once the water enters the atmosphere, it is available again to
rain on the land, dissolving more salts from the rocks.

You can see that, after millions of years, this cycle would
work to continually increase the amount of dissolved salts in
the ocean.

The other process that delivers salts to the ocean is something we discovered only recently, in the last twenty-five
years, in fact. You have probably heard about the deep-sea
hot-water vents, with their colonies of strange animals, that
have been discovered by robotic submarines. These vents are

also a source of salt in the ocean. The hot water coming out of the vents is actually pretty good at dissolving sub-sea rocks, just the way rain dissolves surface rocks. So these vents actually chew away at the sea floor, putting large amounts of salts into solution in the ocean.

You might think, quite logically, that all this salt pouring into the oceans would tend to make the ocean saltier and saltier over time. Fortunately, that is not the case. Over geological time, the ocean has come to a kind of equilibrium with respect to its salt content. Chemical reactions work to limit the extent to which the ocean gets saltier. The chemical reaction known as precipitation is part of the story. When dissolved salts reach a certain concentration, they just re-mineralize, or precipitate, and drop to the sea floor as sediment. But there's also a strong biological drive to take salts from the ocean. For example, some ocean creatures build skeletons or shells from dissolved salts, such as calcium and carbonate, which are combined to form the compound calcium carbonate (chalk), and they acquire these constituents from sea water. So things like clams and plankton are constantly taking up the salts in the ocean and keeping it from getting too salty.

The other question was about salt lakes. There are two kinds of salt lake. One is the kind you see in Utah, where a body of water has no drainage elsewhere, and there is considerable evaporation of its water. Basically these are just like small oceans. Water flows into them, carrying salts, and then evaporates, leaving the salts behind. Over time the lakes get saltier and saltier. There are also some rare lakes that are in geological areas where the surrounding rocks are very salty. In this case, it is not a matter of slow weathering and accumulation of salts.

Instead, every rainfall leaches quite a lot of salt out of rocks in the watershed, and so the lakes end up quite salty. You'll find a few lakes like this in the interior of British Columbia.

NEVER EAT ANYTHING BIGGER
THAN YOUR HEAD
▲

How did small-headed dinosaurs eat enough to sustain them? Comparing the body mass of a Brachiosaurus *with its tiny head, how was the creature able to ingest the great mass of vegetable matter required to nourish its vast bulk?*

Dr. Philip Currie, Curator of Dinosaurs at the Royal
Tyrrell Museum of Palaeontology in Drumheller, Alberta:

For those readers who didn't know, a *Brachiosaurus* is one of the sauropods, and the sauropods were gigantic dinosaurs. Sauropods were the biggest land animals that ever lived, and *Brachiosaurus* was one of the biggest of the big ones. It was as tall as a giraffe, with a very long neck that would stretch up into the trees, but it was twenty metres long, and could reach ten metres high. They probably weighed between forty and eighty metric tonnes, and yet they had long, slim necks and very, very small heads.

I have to admit that paleontologists too have wondered how brachiosaurs could stuff enough food through that tiny head to feed that huge body. Structurally, they had to have a small head, because their long necks could not support any-thing heavier. But they had some evolutionary adaptations that allowed them to move some of the functions that modern animals perform in their heads to elsewhere in their bodies. For example, sauropods didn't really have any teeth adapted for chewing vegetation. The teeth they had actually acted more like the teeth of a comb or rake.

Sauropods would use their teeth to rake leaves off branches and swallow the twigs and leaves whole. The vegetation would go down the neck into the body, and inside the body they had a kind of gizzard – a false stomach filled with stones that would grind up the vegetation. We've actually found the stomach stones of these animals. By moving this function down into the body, they didn't need to have a head big enough to contain large chewing teeth. In fact, if you look at a sauropod head, it is effectively only the brain, the eyes, and the nostrils wrapped around the throat. So as long as the throat was big enough to take in enough food, then the size of the head didn't matter much.

The other question here is just how much food these animals would have needed. That is still a bit controversial because the answer depends on whether these animals were warm-blooded or cold-blooded. Warm-blooded animals need to eat a lot more food, because they maintain their body temperature at a high level all the time. This gives certain advantages in terms of tolerance to temperature extremes, and maintaining activity levels. Cold-blooded animals need less food because they allow their internal temperatures to vary with the environment, but the cost is that they can't always maintain high levels of activity – they can't be quick off the mark. Today we know, for example, that lions need a lot more food than boa constrictors of the same body weight, and it all has to do with the difference between the warm-blooded lion and the cold-blooded reptile.

If sauropods were cold-blooded, they would not have had to eat very much, perhaps a quarter of what a warm-blooded animal of the same size would need. Then their head and mouth size would not be an issue. They would have to eat about the same amount of food as an elephant eats in a day, and that would not be a problem.

If they were warm-blooded, however, it is a different story. Some of my colleagues have calculated that a warm-blooded sauropod would have to eat about half a tonne of food a day, and spend sixteen to twenty hours a day feeding. Some have suggested that this proves they must have been cold-blooded, but there are some animals even today that spend that much time feeding. It is certainly not impossible.

Surprisingly, their size actually gave the sauropods an advantage that makes functional warm-bloodedness a little more likely. Small, warm-blooded creatures have to eat an awful lot to maintain their body temperature because their mass is so small compared to their surface area. They radiate heat away very quickly, and so they need a lot of fuel to stoke up. Mice, for example, need to eat close to their body weight in food every day. Elephants, on the other hand, have a very large mass compared to their surface area, and so they cool much less quickly. They don't eat anywhere near their body weight in food. The larger the sauropod, then, the relatively smaller fraction of its body weight it would need to eat every day.

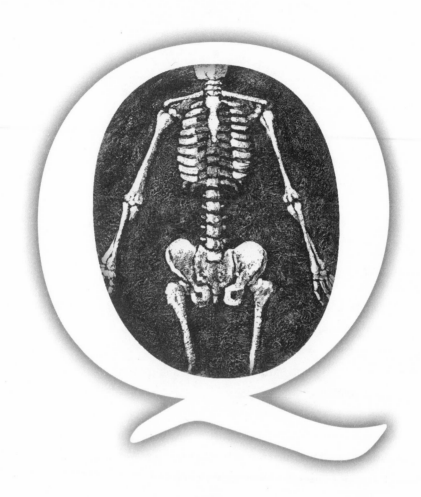

2

Brains, Bones, and Behaviour:

PHYSIOLOGY AND THE HUMAN BODY

You Make Me Boil

▲

What would happen to a human body if it were tossed out into open space? Would it really explode into a million pieces?

Colonel Chris Hadfield, Astronaut, and the first
Canadian to walk in space:

We astronauts think about this problem whenever we go outside a spaceship in our big, white spacewalking suit, the Extra-Vehicular Mobility Unit. It is just a cloth between you and the vacuum of space, so you have to be prepared for popping a little hole in a glove or something worse. Of course, we try to avoid it; but if we do pop a little hole in the suit, there is enough pressurized oxygen inside it to feed the leak for a long time. So we would never be exposed to the vacuum.

But if we were exposed to space, then we would have to worry about the drop in pressure. Here on Earth, at sea level, the air is squeezing our bodies at a pressure of about one kilogram per square centimetre. The higher you go, the less pressure there is squeezing your body. If you go out into space, there is no pressure at all. You become, basically, a bag of skin containing all the fluids and some of the gases that are inside. If you were suddenly released from all the pressure of the atmosphere, then certain things would start to happen fairly quickly. In the movies, they tend to exaggerate the effect, but as the pressure of the air is released, all the gases that are dissolved in your blood are going to bubble out.

It is like boiling water. A pot of water at sea level will boil at 100 degrees Celsius. But take that same pot of water up a mountain and you can boil it at a much lower temperature. That is because pressure keeps the gases dissolved in the water. As you lower the pressure, it is easier for the gases to escape from the liquid, and in space, the pressure is almost non-existent. So, take a human into the vacuum of space and all the gases in the blood and other fluids would start trying to bubble out. While the body wouldn't explode, the fluids would start to form bubbles in the blood vessels and this would cause the damage. You'd have a fatal case of the bends, and you'd be done for.

However, you wouldn't die right way. It is conceivable you could survive the vacuum very briefly. When people have their hands exposed to a vacuum in a glove box, the blood goes to all the little capillaries and the hand turns red. They run the risk of putting a bubble into their blood, which would give them the bends. So you could probably survive a few seconds in a vacuum if you had to. But it is sure not something I want to try.

MAKING YOUR MARK

▲

How are fingerprints formed? Do identical twins have identical prints?

Dr. Clarke Fraser, Professor Emeritus of Human
Genetics at McGill University:

A fetus will get its fingerprints around four months into pregnancy. The patterns are probably related to the tensions of the skin when the ridges are forming, and there is some relationship to the shape of the underlying finger. For instance, if the fetus has a thin, flat finger, then it is probably going to develop the fingerprint pattern known as an arch. Arches are lines running straight across the fingertip. If the fetus's finger has a big bump, a sort of bulbous tip, then it is going to develop a whorl pattern. That is where the lines form into a closed circle. In between, we have the loops. So the patterns are genetically determined, but in a complex way, so you can't identify the particular genes.

If you want to examine these genetic effects, one good way is to look at identical twins. They share the same genes, and their fingerprints are very similar to one another's. In fact, they're just about as similar as those on your left and right hands. They're so similar you can use the fingerprints to establish whether two individuals are really identical twins, with a pretty high degree of reliability. Looking at fingerprints is not quite as good as looking at DNA, but it is still pretty accurate.

There will be some differences between the fingerprints of twins. The differences will be random variations, because fingerprint formation is also affected by the position of the hands while they're developing. Changing the hand position will change the tension on the skin, which will change the forming fingerprint.

Once fingerprints are set, they stay the same for life. They get bigger but the patterns don't change. We might get little nicks and scratches that scar them, but if you wanted to change them completely, that would be very expensive.

OVULATION ROTATION
▲

Humans have two ovaries. Does each ovary ovulate every month, or do they take turns? What controls this, and what would happen if one ovary were removed?

Dr. Jerilynn Prior, Professor of Endocrinology
at the University of British Columbia:

Usually only one egg is produced at a time, but it is not a simple process of the left taking one turn and the right taking the next. In fact, it appears that women usually have more follicles that ovulate from the right side. Follicles are the little nests of cells that surround each egg. It is not clear why the right side would be favoured, but it could have something to do with the blood supply or the temperature.

To stop humans from producing litters, hormones tightly control this whole process. As the follicles mature, one of them gets ahead of the others and starts producing hormones that prevent the other follicles from growing. One of these hormones is estrogen, and another is inhibin (which is really a family of hormones we don't completely understand). Inhibins, as their name suggests, stop the development of other follicles, primarily through inhibiting the production of a follicle-stimulating hormone from the pituitary. The follicle that makes the most hormones gets to produce the egg that eventually leaves the ovary, potentially to get fertilized.

Fertility treatments can override this system, allowing more eggs to be produced for potential fertilization.

When the system is working normally, having two ovaries is a kind of fail-safe mechanism to make sure there is an adequate supply of eggs for a woman's lifetime. It also ensures a supply of the follicles that make the hormones estrogen and progesterone, hormones that we need for optimal health. But if one ovary is removed, a woman usually has perfectly normal fertility and a perfectly normal reproductive life, with menopause at a usual age.

TUNING IN TO A TONAL TRANSITION

▲

Why, as we get older, do our voices change so much? It is not just the child to adult change that concerns me, but the adult to the elderly. What causes our voices to deteriorate?

Dr. Françoise Chagnon, Assistant Professor of
Otolaryngology at McGill University:

Let's start with the voice changes at puberty. The changes are
more evident in young men, of course, but they do occur, to a
certain degree, in young women, too. This is due to the growth
of the voice box (also called the larynx), which is heavily influenced by hormonal changes. The male hormone, in particular,
has a greater influence on the growth of the voice box and
development of the Adam's apple. The entire voice box
and throat are growing and lengthening. Our necks get longer
between childhood and adulthood. And the dimensions of the
throat also increase, making the vocal cords longer and a bit
thicker. The muscles of the vocal cords get stronger and eventually this leads to the mutation of the voice in a young man.
Women also have some growth hormone–mediated changes
to their voice box, but it is a little subtler.

How fast these changes happen varies from person to
person. Sometimes they're quite dramatic.

Like growth spurts in other parts of the body, there can be
growth spurts in the voice box. It can be difficult to adapt
rapidly to the change in mechanics and, for some, this leads
to the voice "breaking." But eventually the young man recov-
ers from that and adapts normally. Once we're adults, our
voices don't change that much, and we retain our vocal identity throughout our lifetime. There are certain characteristics
of our voices that identify us, and those will always remain. It
is true that, as we get on in age, there will be some natural
changes in the voice. This is related to some loss of muscular
strength in the throat, and maybe some loss of pulmonary, or

lung, function. Remember that for a good strong voice, you need a good set of lungs too. You need to push the air out across the vocal cords. Some of that you lose with age.

A small minority of people, with advanced age, will show a significant change in their voice. In these cases, there has been a loss of muscular strength and an atrophy of the ligaments and muscles of the voice box. Some people seem more susceptible to that than others, and we wonder if there are hereditary tendencies for the condition. There are some family trends that support that. But for many others, their voices will remain strong up into their eighties and nineties.

YOU SAY YOU WANT AN EVOLUTION

▲

I have always wondered, when looking at drawings of early humans, what will we look like millions of years from now if we keep evolving?

Dr. David Begun, Professor of Anthropology
at the University of Toronto:

Humans did evolve from something that looked like an ape, and, as paleoanthropologists, my colleagues and I usually try to reconstruct the past. But it is kind of fun to speculate about what might happen in the evolutionary future. Keep in mind that we understand that evolution is an organism's response to a changing environment. So, for humans to continue evolving, we would need to find ourselves in a pretty different environment. One major change that might occur is that we move permanently into space. That would represent a new environmental circumstance to which humans would almost certainly adapt and evolve.

In space we would develop larger chests. It would be advantageous to have a larger heart, perhaps a larger lung capacity. Also, limbs are biologically costly. If we don't load our limbs, or if we don't walk on them or transmit weight through them, they atrophy. And that could be selectively disadvantageous. So people with smaller limbs may be selected for, too. The situation is similar to animals living in the oceans, the low-gravity environments here on Earth. Dolphins and whales both evolved from mammals with limbs that returned to the ocean. They also have enlarged hearts and lungs.

But not everyone from Earth is going to move into space. Those who stay behind probably won't change too much. People are fond of speculating that, because we have noticed a change in brain size over the last five hundred thousand years, our brains are going to continue to grow. But we have probably reached the limit of brain-size increase. Brains are very expensive to maintain as they are, and a brain any larger would become biologically impossible. Then there is the tricky issue of giving birth. If we had larger brains as adults, we would have larger brains as babies, meaning even more difficult births. So it is probably not possible to get brains much larger than we have now.

There is another reason we aren't likely to change. Evolution happens as a response to changes in the environment. But we have such control over our environment today that we can manipulate it to the point where it is not going to have the kind of effect needed for significant evolutionary change.

I'M NOT LISTENING

▲

Why does a snorer never hear him or herself? The noise can be truly awesome, as any victim can attest, and I'm at a loss as to how anyone can sleep through the echoing thunder coming from the pillow next to our own. Why are we the only ones to hear it? Why can't the snorers?

Dr. Meir Kryger, Professor of Medicine at the University of Manitoba and Director of the Sleep Disorders Centre at the St. Boniface Hospital Research Centre in Winnipeg:

This listener is going through something that many people go through, and it can certainly seem mystifying that a snorer can sleep through the tremendous racket they make. The explanation lies, of course, in our brains. Our brain basically ignores information that it doesn't consider important. The snorer's brain just decides that the noise from its own snoring is not going to wake it up.

We used to think that the sleeping brain was basically in neutral, idling and not doing much. All sorts of experiments have proved that this isn't the case, that the brain is extremely active during sleep. We know that it is "hearing," but that it does a lot of filtering and signal processing, so that it will only respond to the kinds of sounds it knows are important. Mothers, for example, become sensitive to the softest cries of their babies when they need to be fed, but will ignore the louder noise of an airplane flying overhead at four in the morning.

A lot of this filtering is going on in a region of the brain called the thalamus. We don't understand the mechanism, though, and we certainly don't understand how the brain decides what's important and what's not important. It is a quite amazing ability. Snorers can reach eighty decibels, louder than a barking dog, and sleep through the whole thing. When you play them a tape of themselves snoring, they can't believe they are able to sleep through it.

Of course, an interesting question is, if the snorer's brain can ignore the noise, then why can't the brain of the person lying next to them do the same thing? The answer is that it can and often does. Snoring is very common, especially in Canada, it seems. In a study done in Toronto several years ago, about 80 per cent of wives claimed that their husbands snored. Other studies from around the world have found that about 30 per cent of adult men and about 15 per cent of women snore. There would be a lot of sleep-deprived women and men out there if they could not adapt to their partner's snoring. Studies have shown that if the bed partner can fall asleep before the snorer, she or he will very often sleep through the snoring. However, the bed partner who doesn't fall asleep before the snoring spouse may be doomed for the night, since the unconscious mind seems to be better than the conscious mind at filtering out that infernal racket.

RISING FROM THE DEPTHS

▲

There is a theory that the baker on the Titanic *survived for two hours
in the freezing cold water because he drank large amounts of whisky.
But most information on alcohol says it makes you feel warmer, but
actually lowers your body temperature. What's going on here?*

Dr. John Hayward, Professor Emeritus of Biology
at the University of Victoria, and an expert on
cold-water survival:

What's going on is a misinterpretation of evidence and some
suppositions that aren't valid. From scientific studies, there is
just no evidence that consuming alcohol will increase survival
time in cold water. In fact, in high doses, such as the baker may
have taken, alcohol will make people cool faster, because
they're slightly anaesthetized and their shivering and other
thermal defences would be blunted. So there is no evidence
that alcohol would have allowed the baker to survive.

What was probably important for the baker's survival
was his body build. Science has shown that the main deter-
minants of how long you can live in cold water are fatness and
body size. These two factors increase the body's insulation
considerably. As a generalization, if you double the amount of
fat under your skin, you double your survival time. It is easy
to imagine that the baker, if he were middle-aged, would have
eaten more of his own baked goods than he should have, and
he would have been fat. That is probably why he survived: his
own blubber would have saved him, just like the blubber

on a whale or seal prevents it from getting hypothermia.

Also, the baker's body size could have helped keep him alive. A child in the water cools down fast, because it has a lot of heat-losing surface area compared to its heat-conserving and -producing mass. An adult, such as a well-muscled, heavy-boned person, independent of how much fat he has, has a much slower cooling rate because his surface area is relatively small compared to his volume.

So the combination of being about ninety kilograms and 25 per cent fat, which is common in many middle-aged male adults, greatly increases survival time. The average adult can survive about one hour in icy water, but this can be doubled or tripled with obesity and large size. So it is easy to speculate that this is how the baker avoided a fatal level of hypothermia. Also, if he had been able to raise even a small portion of his upper body out of the water while hanging on to some floating debris, this would have significantly slowed his cooling rate. Of course, there could have been a problem for the baker if he had been drinking, but it would be more of a problem after he got out of the water. At that point, he would need good shivering to produce enough heat to help warm his body back up. And alcohol, even in low doses, blunts shivering.

Finally, in answer to the listener's specific question, "What's going on here?" it must be understood that, although the tranquillizing effect of alcohol can make your skin feel warmer, this is only when you're in the usual, comfortable room temperature. Obviously, this wouldn't happen under cold stress, such as immersion in frigid water. Believe me, neither the baker nor any other of the victims in the *Titanic* tragedy felt warm!

TREMOR STEMMER

▲

*When a person has Parkinson's disease, there is an involuntary,
uncontrollable tremor. Why do these tremors cease when they sleep?
What makes them stop?*

> Dr. Janis Miyasaki, Neurologist at the University of
> Toronto, and Associate Clinical Director of the Movement
> Disorders Research Centre of Toronto Western Hospital:

The reduction of tremors in sleeping Parkinson's patients is
a reflection of normal sleep patterns. When we go into stage-
four sleep, a very deep form of sleeping that occurs when
we're dreaming, it is part of our normal makeup that we can't
move. If you think about it, this makes sense. If we were acting
out our dreams, it would make for a very disruptive night. So
while we're dreaming, our muscles are completely flaccid, or
floppy, and we can dream in our minds and be quite still in
bed. It is a sort of paralysis.

In Parkinson's disease, the tremors are involuntary and
start in a part of the brain called the thalamus. The thalamus
is a pacemaker for the tremors: it sends out regular signals to
the hands or limbs, telling them to shake. During deep sleep,
these signals are still being sent out, but the relaxation of the
muscles overrides the instructions, allowing the Parkinson's
sufferer to sleep.

As sleep lightens, and the Parkinson's sufferer starts to
come into other stages of sleep, the tremor might actually wake
them up. So Parkinson's patients frequently have difficulties

falling asleep, and their sleep is more disrupted. They also have difficulties falling asleep again if they wake up in the middle of the night.

Interestingly, there is a possible direct connection between this disrupted sleep and Parkinson's disease. The same chemical that is missing from the brains of Parkinson's patients, dopamine, is very important in a part of the brain that regulates sleep and regulates our internal or biological clock. There are many different medications to relieve the symptoms of Parkinson's disease. They all act to replace dopamine and restore chemical balance in the brain.

PRIMATES' PROGRESS
▲

If evolution is the underlying principle of creation, why don't apes still evolve into men?

Dr. Mary Pavelka, Associate Professor of Anthropology
at the University of Calgary:

There is a common misconception that humans have evolved from apes. Equally logical questions would be why humans haven't evolved into apes or why lions have not evolved into tigers, or dogs into cats. Apes have their own separate evolutionary history. They evolved into apes at the same time that humans were evolving into humans. So apes are, in fact, contemporary relatives of ours, not our ancestors.

Part of this misconception is based on the idea that evolution is some kind of directed process that led to humans as the ultimate goal of the process. But biological evolution is not progressive, it is not directed, and there is no purpose to it. So this popular idea of humans as the ultimate evolutionary success story is not accurate.

Great apes do not represent an earlier stage through which we passed. We share a common ancestor with them, a species that lived some seven or eight million years ago. Since the time of that common ancestor, we have each been on separate evolutionary paths heading towards the living descendants you see today. So the living descendants of that ancestor on the great-ape line, the animals we share the planet with today, cannot be our ancestors. None of the living apes today will ever evolve into humans. Apes are perfectly evolved apes, and we are perfectly evolved humans, and there is no reason to assume that an ape should be trying to evolve in the direction we took, nor that we should be trying to evolve in the direction that apes took.

The common ancestor for both humans and apes was probably more ape-like than human-like. It was, in all likelihood, a quadrupedal animal (one that walked on all fours), and possibly an arboreal, forest-dwelling animal. We don't know as much about the common ancestor as we would like to, probably because it was a forest-dwelling species, and conditions in the forest are not really suitable for fossilization. Fossils in the forest are quite rare.

One reason for assuming that a common ancestor to humans and apes must exist is how biologically similar we all are. We can get some of the same diseases; leukemia, for instance. And we have the same ABO and Rh blood groups. From what we have mapped out so far of their genomes and ours, we estimate 99 per cent similarity in our genetic material. We're very closely related. Culture change has made us very different in many respects, but biologically we have the same basic template, and are not nearly as different as many people tend to think.

It is highly unlikely that apes could evolve into a more human-like form. The chance of the evolutionary process replicating the same species twice is so infinitesimally small that we can safely say it is impossible. So we shouldn't expect to see some *Planet of the Apes* scenario, where apes have become human, and only retain some of their body hair. There is no reason to assume that would happen.

FEELING BLUE, GETTING RED
▲

Since the normal environment for our eyes is to be moist with tears, why do they get bloodshot when we cry?

Dr. Charmaine Chang, Ophthalmology Resident
at the University of Alberta:

First and foremost, there are two types of tearing. There is background or basic tearing, and the other is reflex tearing.

Every day, twenty-four hours a day, our eyes produce tears. However, this is a different sort of tear than we produce when we cry or get something in our eye that is irritating.

The composition of the two types of tears is different. Background or basic tears have an oily component, a watery component, and a mucous component. They're designed to keep the eye moist. In contrast, reflex tearing is a more watery or aqueous type of tear. The other main difference is the source of the tear. Reflex tearing comes from the lachrymal gland, which sits beside our eyeball, in the outer upper corner of the eye socket. Background tearing comes from glands that sit in the underside of our eyelids, and in the surface, or conjunctiva, of our eye.

So, the two types of tearing are quite different, and only the reflex tearing causes our eyes to go red. That is because reflex tearing is an automatic response that starts up a number of different effects. Nerves that supply the lachrymal gland stimulate this tear production. These nerves also enervate the blood vessels that supply the eye. When we cry the signals from these nerves cause dilations of these blood vessels, and we see it as redness in the eye.

You can think of this as a defence mechanism for the eye. If something is irritating or harmful and approaches your eye, or is actually in your eye, the tears provide protection. The tears contain various enzymes and chemicals that help prevent infection, and the tears themselves can flush out whatever is in the eye. At the same time, when you bring more blood to the surface of the eye, there are protective chemicals and proteins in the blood that can provide a defence as well. Why emotions can make us cry and what role tears play then is not really clear. The inputs that drive emotional tearing come from higher centres in the brain that we don't understand well.

Of course, it doesn't help when we rub our eyes. That just increases the redness that started with the crying.

MIND OVER MARATHON
▲

A marathon runner runs all day until he's exhausted. That same day, a particle physicist is presented with an interesting and difficult problem. She sits at her desk wracking her brain to solve the problem. Which one is more exhausted at the end of the day?

Dr. Max Cynader, Director of the Brain Research Centre
at the University of British Columbia:

The answer is: they are both exhausted, but in different ways. Running probably uses more energy, if you think of just the caloric requirements of having to move a seventy-kilogram mass. But we do know that pathways in the brain that are used to excess tend to fatigue and show what is called adaptation. You see this adaptation at a lot of different levels. One of the simplest ways to look at it is with a sensory system. If you exercise a particular pathway for a while, you can see it getting tired. For instance, if you look at a waterfall, with the water going down all the time, after a while the visual response to that motion diminishes and you won't notice the

downward-falling water. If you then look at a neutral stimulus, like an ordinary wall, you will see things going upward,
because of the weakening of the downward-direction sensing
system, which occurred during the adaptation process.

There is a good chemical reason behind this. In the case of
the particle physicist, she is using a set of pathways that may
be very specific to solve this problem. Brain cells are made
from a cell body and a long wire called an axon, which con-
nects to the next cell. Brain signals pass from one axon to the
next cell by the transfer of neurotransmitters. The neuro-
transmitter used for most long pathways involved in active
thinking is called glutamate. There is evidence that if you use
the same pathway for a long time, you will release less glutamate later than when you first started. You will actually run
out of reserves of glutamate, just like a car running out of gas.
If you have less of this transmitter, it is less able to excite the
next cell, and the whole pathway is weakened.

Interestingly, in the long term, pathways that are used
frequently can actually become stronger. In that sense, they
are similar to muscles, which, when exercised, become tired
in the short term, but in the longer term hypertrophy and
become stronger. There is evidence that if you spend a lot of
your time thinking about particle physics, your ability to think
about particle physics will improve, and that would be partly
because the pathways that are used over time become stronger.

But as far as comparing runners to thinkers, that is not
really a fair comparison. We are really talking about apples
and oranges.

BITTEN BY THE COLD

▲

Why, when you've been frostbitten, is that body part now more sensitive to the cold, even months or years later? I would have thought there'd be some sort of nerve damage that would reduce sensitivity afterwards.

Dr. Gordon Giesbrecht, Professor in the Faculty of
Physical Education and Recreation Studies, and a
hypothermia expert at the University of Manitoba:

The second part of the question is certainly correct. If you kill an area by freezing it, you usually lose that part of the body, and you wouldn't feel anything. But in the parts that survive, and are not amputated, there may be nerve endings that are damaged. These may get confused and give a multitude of different kinds of responses to stimuli.

There are two types of nerve endings in the skin, temperature-sensitive nerve endings and pain nerve endings. Either may recover over a period of time, or they may never recover. Nerves do regenerate, as long as they're in tissue that is oxygenated. So as long as the capillaries aren't completely destroyed, oxygen can still be delivered, the tissue will stay alive, and the nerves can regenerate in that tissue area. When these nerve endings are damaged, they may become more sensitive. They are not operating properly, so the response from a given cold receptor or pain receptor to a stimulus might be greater than you'd normally get.

As well as messed-up pain receptors, there is a second problem that happens after frostbite. Although the blood vessels under the skin haven't been destroyed, they have been damaged. In a subsequent cold stress, the blood-vessel walls are more sensitive to cold and actually constrict faster. Normally, when we get cold, our blood vessels in the periphery constrict, in order to try to save heat for the core of the body. But after we have been frostbitten, the vasculature will constrict sooner and even more than it normally would. Therefore, the damaged tissue is actually getting colder than before because there is even less blood flow.

So, getting hit by frostbite is really a double blow. The first thing is that the damaged area is more likely to become frozen the next time you are out in the cold. The second thing is that the neural sensitivity is increased, or at least it is changed. Once you've been frostbitten, your tissue is more susceptible to frostbite and will also be more sensitive to the cold itself.

SEEKING SINISTER SOURCES
▲

What makes someone right- or left-handed? Why are most people right-handed?

Dr. Murray Schwartz, Associate Professor in Psychology
at Dalhousie University:

For the most part, our handedness is genetically determined, but it is not clear whether it is one pair of genes or two pairs of genes. There are also lots of other influences that go on. We seem to inherit our handedness from our parents, but not in a strict fashion. If you have two left-handed parents, you are more likely to be left-handed than people who have two right-handed parents, but it is certainly not guaranteed. Not all left-handed parents have left-handed children.

In addition to genetics, there are tremendous early environmental influences that affect development. Prenatal influences, such as temperature and hormones, can affect handedness. As we find out more about brain chemistry, we're finding subtle influences there too. But we're not at the point where we can say *this* is the gene for handedness, and if you have this, it causes right-handedness, and if you don't have it, it causes left-handedness.

Even more intriguing is how consistent handedness has been throughout history. The amazing thing is that the proportion of left-handers in the population, which is believed to be about 10 per cent, seems to have existed for as long as we have had recorded history. There are all sorts of archaeological finds where about 9 or 10 per cent of the tools used by prehistoric people seem to be left-handed.

We talk about left-handedness and right-handedness as if they are two ends of a pole, but in fact handedness is much more of a continuum. If you are right-handed, the chances are you'll do almost everything with your right hand. But if you are left-handed, you are much more of a mixed bag. There are not that many left-handers who are exclusively left-handed, because many of them have learned to do things with their right hands. They have acquired a certain amount of dexterity, which is the appropriate word, since dexterous comes from the Latin for right, *dextera*.

This flexibility seen in left-handed people is reflected in the structure of their brains. People have looked at something called hemispheric specialization. Ninety-six per cent of us have the centre for speech in the left side of our brains. But if you look at left-handers, the number drops to 60 per cent. Forty per cent have a different sort of organization. Of the 40 per cent, half have their speech centre in the right side of the brain and half have their speech function split between the two hemispheres. So it could be that the brains of many left-handed people are organized differently from those of right-handed people.

Handedness seems to be a unique trait of human beings. Other mammals might show a preference for one side or the other, but across a specific species it seems to be much more even. There is more of a fifty-fifty split. It is just in humans where we find this huge species preference for right-handedness.

CATCHING THE COMMON YAWN
▲

Why do we yawn and why is yawning contagious?

Dr. Irvin Mayers, Professor of Medicine at the
University of Alberta:

The short answer to this question is, nobody really knows. There is a whole variety of myths related to yawning. Some suggest it is a form of aggression, baring teeth. Others have suggested that yawning improves oxygen levels or improves carbon dioxide levels. But the one thing we do know is that there has been a variety of studies that have looked at yawning, and it is very clear that yawning has very little to do with breathing.

There have been studies that have added carbon dioxide to the air people are breathing in, and it did not alter the frequency of yawning. You can also increase the oxygen content without altering the frequency of yawning. In fact, you can still take the same big breath if you yawn through your clenched teeth or through your nose, but the desire to yawn won't be satisfied.

Yawning is a very primitive reflex. Ultrasound studies looking at a developing fetus have shown yawning-like behaviour within the first three months of pregnancy. When we look at the neurobiology of yawning, it seems to be centrally controlled in the brain, and can be affected by changing chemical levels. Drugs like serotonin and dopamine, both of which are involved in transmitting signals in the brain, can increase the frequency of yawning. Opiates decrease the frequency of yawning.

Ultimately, yawning may be more about stretching the face muscles than breathing. It is an arousing type of reflex. It actually wakes you up a bit. There also seems to be a difference between the way we yawn in the morning and the evening yawn. Morning yawns are more associated with a general body stretch, and may help arouse the body. Evening yawning tends to be mainly just stretching the face, getting you ready to settle down for sleep.

Yawns are truly contagious. Again, studies have looked at this. If you look at a picture of someone yawning, you yawn. If you read about people who are yawning, you'll yawn. If you hear a yawning noise, you yawn. You can't catch a viral infection by reading about it, but you can certainly catch a yawn this way. Unfortunately, there is not much research into finding out why people yawn. It would need a lot more work in the areas of psychology and group behaviour, but there is not that much interest in it. A lot of the granting agencies would probably consider the whole field a big yawn.

NEXT STOP, OVIDUCT
▲

What motivates sperm? Do they have some sort of primal brain that programs their movement or are they just influenced by their surroundings?

Dr. David Mortimer, President of Oozoa Biomedical in Vancouver and an international consultant in reproductive biomedicine:

The simple answer is, they've got no idea where they are going. Since a spermatozoon, or sperm cell, is just one single cell, it doesn't have a brain. It is just one cell with its nucleus packaged away, inactive and ready to be delivered to the egg.

Basically, sperm just keep swimming and some are lucky enough to find the egg.

A sperm cell is made up of two parts. The head contains the nuclear material that is destined to combine with the DNA in the egg to create a new person. The tail is a slender whip-like structure called a flagellum. Its job is to propel the sperm towards the egg.

The sperm's movement, or motility, is very important at certain times, such as when they're getting into and through the cervical mucus. That would the first barrier they face after being deposited at the upper end of the vagina. Once they get through the mucus and into the uterus, the fluid lining the uterine walls generally mixes them around. Random contractions of the uterus propel them further up. For this stage, their ability to swim is not important.

Once they get into the oviduct, motility becomes important again. The oviduct is part of the Fallopian tube. Here, a reservoir is established* and the sperm move on farther to the site of fertilization in response to certain hormones released from the ovary just before ovulation. Once they get closer to the egg, some interesting changes to their motility occur. The pattern changes substantially, due to changes in the way the sperm tail is beating. Now the sperm swim harder. If you look at them under a microscope, they look like they're just thrashing around. But in fact, at this point, the head of the sperm is attached to the cells around the egg, and this thrashing motility is what drives it to push through the cell layer around the egg.

---

* The sperm stay in this reservoir, with suppressed motility, until sometime close to ovulation, when some of them are released upwards.

In humans, this thrashing is still random, but in some other species there is a chemical attraction of the sperm to the egg. This is called chemotaxis, and has been observed in some species of fish and invertebrates. But no one has ever actually documented, without questions being raised, a process of chemotaxis in mammalian sperm. That is partly why there are so many of them. They have a large volume to occupy, and some sperm go up the wrong Fallopian tube as well. It is basically just chance that any fertilization ever happens.

ELECTRIFIED ENAMEL
▲

Why do we feel such discomfort when a piece of aluminum foil touches a dental filling?

Dr. Dorin Ruse, Associate Professor of Biomaterials in the Faculty of Dentistry at the University of British Columbia:

My father's a physician, and he says that if something hurts, don't do it. But if you have done what this listener has done, then you should know that the reason you felt discomfort is because of a phenomenon known as galvanic shock. When you touch two dissimilar metals, it is like short-circuiting a battery, and that is exactly what happens when you touch aluminum foil to an existing metal filling in your mouth. In effect, an electric current flows between the two metals, and the nerves in your mouth, which are very sensitive to electrical currents, feel the jolt.

What is actually happening here is that your mouth is acting like a wet-cell battery, known as a galvanic cell. The saliva in your mouth is full of ions – charged particles – and it acts as an electrolyte, like the acid in a battery. The metal in the filling acts as one of the terminals and the metal in the aluminum foil as the other. When you touch the two metals together, you short-circuit the battery and a large current flows.

This will work with other metals that might find their way into your mouth, as well as aluminum foil. For example, if you touch a filling with the tines of a fork, you may well experience it. But not everybody experiences a shock. We have different thresholds to pain. Studies have shown that some people can feel pain from tiny currents – just one to five microamperes. Others can tolerate up to a hundred micro-amperes before they experience pain. In fact, dentists have found that some people with a very low threshold can't toler-ate metallic fillings at all, just because of the small currents they experience. They actually go to the dentist and ask for the filling to be replaced or removed.

For other people, the sensation can decrease with time and is usually only a problem with new fillings. After a while, the nerves near the filling can lose their sensitivity and better withstand these small currents. In order to protect the tooth during the few days when the nerves are most sensitive, some dentists put varnishes over newly placed metallic fillings to insulate them, but it is not a common practice. Generally, you can avoid the problem by just taking the aluminum foil off the gum or candy before you eat it.

A SEASONAL SPASM

▲

Why is it that every time I go out for spicy food – Indian food or Thai food or something like that – I always get the hiccups?

Dr. George Bubenik, Associate Professor of Physiology in the Department of Zoology at the University of Guelph:

This is one of those physiological questions that are not easy to answer, because we don't have all the answers. Hiccups are caused by a spastic contraction of the major respiratory muscle, which is called the diaphragm. It is located between the chest cavity and the abdominal cavity, and it is kind of dome-shaped. When it contracts, it creates a vacuum that sucks air into the lungs. When it releases, the air puffs out. Hiccups are just an uncontrolled contraction of this muscle.

As to what causes hiccups, we know that overstimulation of the upper part of the stomach and the lower part of the esophagus, the tube that leads from the mouth to the stomach, will cause a contraction of the diaphragm. The top part of the stomach goes right through the diaphragm. When you stimulate it by eating large chunks of food, drinking a freshly opened pop with a lot of fizz, or eating very sweet or very spicy food, you can overstimulate the nerves in this area, and hiccups will result.

That overstimulation can happen in a lot of different ways. I suspect that in this case, the spicy food is probably causing changes in the chemistry of the stomach, and may cause a production of gas. As this gas rises in the stomach, it puts pressure on the diaphragm. The diaphragm responds by contracting, and these repeated contractions are what we call hiccups.

Many people are interested in trying to control hiccups, and you probably know of half a dozen old-fashioned remedies. Some of these remedies are directed towards the pressure against the diaphragm. So you can bend over, you can pull your knees towards your chin, or you can try to exhale with your mouth closed. Holding your breath also puts pressure on the diaphragm, but it also just calms you and your overexcited diaphragm. That is also probably the secret behind my grandmother's favourite remedy, which was to place a sugar cube on the top of the tongue and let it melt slowly.

LUNCHTIME LETHARGY

▲

Why do people feel sleepy in the afternoon?

Dr. Stan Coren, Professor of Psychology at the
University of British Columbia:

This is a question I get asked fairly often, and the answer has to do with the natural cycles in our bodies. Everybody knows that every twenty-four hours we have a period of time when we go to sleep, and then we wake up and we have a period of time when we're alert. What most people don't know is that the cycle is really two twelve-hour cycles, not one twenty-four hour cycle. The twelve-hour cycle is arranged so that you become maximally sleepy between one and four a.m. – and then twelve hours later, between one and four p.m. So at that time in the afternoon, the pressure builds to go to sleep.

The interesting thing is that this is a very strong urge, one that has little to do with anything external. Many people suspect that it may have something to do with eating a big lunch, but that is not what the science shows. If you were given no food for breakfast, lunch, or dinner, you would still get sleepy between one and four p.m. Many people suggest that heat can sometimes induce the desire for an afternoon siesta. But people are just as sleepy at one p.m. in Edmonton in winter as in Mexico in summer. So it has nothing to do with the afternoon heat.

We should keep in mind that these two twelve-hour cycles are not exactly the same, because obviously we are built to sleep through the hours of darkness, and stay awake through most of the day. In terms of evolution, we were really designed to sleep between nine and ten hours out of every twenty-four, with a big bite in the dark hours, probably about eight hours or so, and a little bite in the afternoon, about one to two hours. That is what most of our primate relatives – the monkeys, the gorillas and so on – do. There is also some evidence that if you do take a siesta, it can be good for you. A study was conducted in Greece, where some people still take an afternoon nap and other people are trying to be more modern and have given it up. It looked at men between fifty-five and sixty-five years of age. They found that the men who were still taking an afternoon nap had considerably fewer heart problems. It may well be a practice we should all think about getting back to.

DOUBLE THE TROUBLE

▲

Is it possible to have the same cold more than once, and is it possible to have more than one cold at a time?

Dr. Kathryn Wright, Assistant Professor in the Department of Biochemistry, Microbiology and Immunology at the University of Ottawa:

You can, in fact, get the same cold more than once. There are about 150 different strains of the two kinds of viruses responsible for common colds: the rhinoviruses and corona viruses. Every time you get any one of these strains, you develop an immunity to it, but it only lasts about two years. So you can get the same cold virus twice. You just can't get it until after your two-year immunity is gone.

There isn't a lot of research on getting more than one cold at a time. However, a study that was conducted over twenty years ago indicated that you can become infected with more than one rhinovirus at a time. But it only occurred in about

4 per cent of people in the study. You can also become infected with a rhinovirus and a corona virus, or a rhinovirus and an influenza virus, which often causes colds in children. That also happens no more than 5 per cent of the time.

Part of the reason this doesn't happen more often has to do with the way the viruses spread. Viruses tend to circulate in communities at different times of the year, so it is probably rare to be exposed to more than one virus at a time. There also is an immunological reaction that prevents a second virus from getting in, once the first has taken hold. Two to four days after infection with a virus, infected cells start producing a group of proteins called interferons. These interferons can then protect neighbouring cells from becoming infected with a virus. So, once you become infected with any virus, you are somewhat protected from infection with a second virus, at least for a short period.

In summary, it is possible to get two colds at the same time, but it doesn't happen very often.

HEAD IN THE STARS
▲

In the comics and cartoons, when somebody has been hit on the head and is a bit woozy, it is usually depicted as the person seeing stars. I have experienced this a couple of times when I'm dizzy and about to faint. I see little silver slivers similar to shooting stars. What am I actually seeing?

Dr. Barry Beyerstein, Professor in the Brain Behaviour
Laboratory and the Department of Psychology at
Simon Fraser University:

The technical name for these perceptions is phosphene, and
a phosphene is any visual sensation that is not initiated by light
hitting the retina. The source of this phenomenon lies in the
fact that the brain can't really tell what is activating its neural
cells. It just knows that the visual pathway to the brain has
been activated somewhere along the route. The brain interprets that activation as though it were caused by whatever
normally activates those cells. Thus, in this case, the person
experiences it as light.

The normal activation for cells in the visual cortex comes
from an impulse from the optic nerve, which is triggered by
light hitting the receptor cells in the eye. But there are a
number of other ways those cells in the visual cortex can be
activated. In the case this listener describes, he said he was
feeling woozy and close to fainting. My guess is that maybe he
stood up too fast and, for a moment, there was a little less
blood going to his brain than is normal. That may have caused
some disruption in the visual cortex, especially in inhibitory
nerve cells that hold the others in check. This temporary
release from inhibition may have allowed some neurons to
fire spontaneously; for example, the ones that normally signal
the presence of light. This could well have been interpreted as
stars in front of his eyes. Temporary interruption of blood flow
to the visual cortex during the initial stages of a migraine also
produces phosphenes that look like shooting stars.

The same thing can happen with physical stimulation of the nerves. If you get a good whack on the back of the skull, where the visual cortex is located, the mechanical energy from that physical blow will activate some neural cells, and so once again you'll see stars. You see something visual that really has nothing to do with your eyes at all. It is the same phenomenon as when striking your so-called funny bone near your elbow causes you to feel that annoying tingling in your fingertips. In that case, the nerve pathway supplying the fingertips has been activated partway up to the brain. The "finger" part of the brain doesn't know this, and interprets the message as having come from the fingertips.

The phenomenon is quite well understood by physiologists, and it is called Müller's Doctrine of Specific Nerve Energies, after a German physiologist in the last century who did some pioneering work.

Dr. Johannes Müller was quite well known in his day, and was even an expert witness in a famous trial in which just this phenomenon came up. In that case, a witness had identified an accused criminal based on a sighting in a very dark room. The defence lawyer asked him how he could be sure of the identity of the accused, given the darkness. The witness said that he'd been staring in the direction of the accused man when one of his accomplices snuck around behind the witness and struck him on the back of the head. In the glaring flash of light at the impact, he'd seen the face of the accused. Professor Müller was brought in as an expert witness to explain that, just because this person experienced that blow on the head as an immense flash of light didn't mean that there was any light whatsoever in the room. It was, in fact, only in the witness's brain that this event had occurred.

A WALK IN THE DARK
▲

Why do sleepwalkers walk?

> Dr. Jeffrey Lipsitz, Medical Director of the Sleep
> Disorders Centre of Metropolitan Toronto:

The truth is, we don't really know. We've studied sleep-walking a lot, and we know that it is a very common problem, especially among children. As many as 20 to 30 per cent of children, at some point in their lives, have experienced a sleepwalking episode. It could be something as trivial as getting up in a semi-sleeping state to go to the bathroom, walk around the house, have a snack, or even peek under the Christmas tree, and then getting back into bed and having no real awareness or recollection of what's happened to them.

Sleepwalking seems to happen during a certain deep stage of sleep that typically occurs within the first third of the night, and it seems to be a partial disturbance of that very deep stage of sleep. It is not the dreaming stage of sleep, where the brain is very active, and the eyes are moving under the lids, in what is known as REM or rapid eye movement sleep. In that stage of sleep, the body is typically paralysed, so that we don't act out dreams. That paralysis is probably a protective mechanism.

The state in which sleepwalking occurs is in the non-dreaming stage of sleep. The brain is active, and the body, obviously, is capable of movement, and goes through familiar tasks. The person often will get himself back into bed, or allow himself to be gently led or pushed back in to bed. Sleepwalkers typically don't do anything dangerous during their nocturnal rambles, so it is relatively safe as long as certain precautions are taken. With children, locking cupboards and doors is probably a good idea, and putting away sharp objects.

You might have also heard that it is dangerous to wake a sleepwalker. That is a bit of a myth. Certainly, when people are awakened from certain sleep stages, they can be quite disoriented and frightened. Children with night terrors experience a similar phenomenon when they can't wake properly, and then can't even recognize the parents who are trying to comfort them, which can be very distressing. The solution is to try to wake them very gently and calmly, and avoid frightening them.

As I mentioned, sleepwalking is much more common in children than in adults. This may be partly related to the development of the brain over the first few years of life. There are a number of sleep disorders that seem to be more prevalent in children, such as bed-wetting, nightmares, night terrors, and sleepwalking. In fact, when we see sleepwalking and certain other sleep disturbances arise out of the blue in an adult who has never had this problem before, we may need to make sure that it is not indicative of some other sinister process going on. In children, though, it is both common and relatively innocuous. Most episodes of sleepwalking pass normally, so it is not something that you are likely to have to put up with for all that long.

Of course, everyone reacts differently to sleep disorders, and people should consult their family doctor if the problem persists or causes difficulties.

I SCREAM FOR ICE CREAM
▲

What causes ice cream headaches? Why does your head ache during or after drinking something really cold, like a crushed-ice drink?

Dr. Jock Murray, Professor of Neurology and Medical
Humanities at Dalhousie University Medical School:

When you experience this unpleasant sensation, what you are doing is cooling the back of your throat and stimulating the nerves there. Those nerves respond by constricting the blood vessels in the face, and the ones that respond most dramatically are just over the eye. That is what gives you the painful sensation of a spike in the eye when you are unwise enough to slurp your ice cream a bit too quickly.

The body's natural response to cold is to produce a narrowing or constriction of blood vessels. We can often see this in our hands. If we put our hands in cold water, or expose them to winter temperatures, they will start to look white. But blood vessels that are constricted for some time start to become painful. The pain is related to the blood vessel tightening further and reducing the amount of blood that can get through.

The interesting thing about these cold headaches is that, while the cooling is in the back of the throat, the nerve response and the pain is in a different area. This is an example of what is called referred pain. In this case, the stimulus in the back of the throat is associated with two large blood-vessel systems, but the vessels above the eye are very small ones, and they're the ones that experience the most severe pain.

Our understanding of this phenomenon came from recognizing an association with another, more serious medical problem – migraine headaches. We were puzzled for a long time about ice cream headaches, but one thing that we noticed was that people who got them also often got migraines. In migraines, stimuli that the body can perceive as harmful, such as stress, bright lights, certain kinds of foods, or paint fumes, trigger a stronger than usual neurological response. One of these responses is that the blood vessels constrict. If there is an exaggerated response, in which the blood vessels overreact, the person often experiences pain, and the pain of a migraine is a throbbing pounding that occurs in the blood vessels in the head. In ice cream headaches, we have a small variation on that by a local stimulus producing a local constriction just above the eye. So your tendency to experience ice cream headaches is linked to the blood vessels that respond easily to external stimuli. It is just an exaggeration of a normal, protective body response.

In any case, the only real treatment for this condition is to avoid the situation that triggers it. Just eat the ice cream slowly and enjoy it.

PISCINE PERSPIRATION

▲

Is it possible to sweat when we swim?

Dr. Norman Kasting, Associate Professor of Physiology
at the University of British Columbia:

I can answer this question from personal experience because
I have spent many hours in training, swimming up and down
in pools, and yes, we do sweat when we swim.

Normally, when you're being active on land – running, for
example – the muscular work you do generates heat within
your body and you have to get rid of that heat. The first thing
the body does is increase the blood flow to our skin. With all
that warm blood flowing close to the skin, we're acting kind
of like a car radiator, attempting to shed heat to the environment. If that doesn't get rid of enough heat, then we start to
sweat. Sweating cools our bodies by putting water on the skin,
which then evaporates, and the evaporation of the water into
vapour cools the skin.

In the swimming pool, this picture is complicated by a
couple of factors. First of all, water conducts heat away from
your body about twenty-five times faster than air does. So it
is very difficult to generate enough heat when you're swimming to increase your body temperature. If you do manage to
work hard enough to increase your temperature, or if the
water is very warm, then you will start to sweat.

It works this way: when you first dive in, your skin
temperature is usually considerably higher than the pool

temperature. Your skin temperature is usually about 28 degrees Celsius and the pool temperature would be, say, 22 degrees. So you dive in, it feels cold. But as you start to work and generate heat in your muscles, that heat will be taken away by the water, because it is a good conductor, and that'll occur for a period. But you'll get to a point where you'll be generating heat faster than the water can take it away. At that point, your body temperature will start to rise and that will trigger the sweating reaction.

The body has the same response to overheating whether it is on land or water. Of course, sweating in water does not help to cool you, because it doesn't evaporate and take the heat away with it. Instead, the sweat just washes off the skin, and we can't even tell we've been sweating.

On the other hand, you might have had the experience in a hot tub or a Jacuzzi where most of your body is submerged, and your head is sweating profusely. In that case, your whole body is actually sweating, but you can only feel it on your head. A way to verify if you have been sweating in the pool is to weigh yourself before and after your workout. If you have a tough workout in a very warm pool, you could discover that you have lost a significant amount of weight through sweat. Of course, regulating our temperature, which we can do so well on land, is a very important part of being able to achieve peak performances for athletes. As a result, serious swimmers have very precise requirements when they swim, and will complain considerably if the water temperature is not almost exactly 25.5 degrees. At 25.5 degrees, the heat is taken out of your body at just about the right rate. That way, you don't get too hot and end up having to sweat, or, on the other hand, you don't get so cold that your muscles are inefficient or subject to something like muscle cramps.

BACK TO THE SOURCE

▲

I remember learning that all of us start our lives as a single cell. I also learned that some of our cells last our entire lifetimes, while other cells die and are replaced by new cells. What I wanted to know is whether that first cell is still in me somewhere, and if it is, I would like to know what it is up to.

Dr. Paul Lasko, Professor and Chair of the
Biology Department at McGill University:

That first cell isn't really there any more. It is not saved up or sequestered away in the body somewhere. In fact, it is divided many, many times and components from that cell have been passed on to the many cells that arose from that original first cell. The way the process works is that once that first cell has divided, and its offspring have divided again, the embryo starts making proteins from its own genes. Those new proteins are what the embryo needs in order to produce different kinds of cells from different parts of that original cell. So by the time the hundreds of billions of cells have come from that original first cell, there's really nothing left that is recognizable as being that first cell. It has been broken down and recycled and distributed throughout the body. So you might, if you were feeling romantic about it, look at most of the cells in your body as having some little remnant of that first cell.

It is true, though, that there are some cells that we form early in development that do survive for nearly our entire lives. The best example of that is in the female ovary, where all the cells that will give rise to eggs, called oogonia, are all produced during fetal development. So a female human has the biggest population of those cells when she's about a five-month-old fetus. Initially, she has several million of those cells, but by the time she's born, it is down to a million. Those cells never divide any more, and they stay there for as long as forty or fifty years. Then, each month, a few of them try to develop into eggs.

That is the best example, but there are other cells that live a very long time. Nerve cells, for example. You have doubtless noticed that babies have rather large heads in proportion to their bodies. That is because just about all the growth in the brain takes place during fetal development and in the first couple of years after birth, and the replacement and growth of new neurons is very, very slow. So some will last a lifetime.

This is a pretty amazing contrast, if you think about it, because quite a lot of our cells last only a matter of weeks or months. There is massive turnover, for example, in the blood and skin, where cells are constantly being replaced by the body.

HEAT A FEVER, COOL A COLD

▲

*I understand that the body increases its temperature as one way of
dealing with foreign viruses or bacteria. What triggers this increase,
and how does the body do it? Then, when the temperature gets too
high, how does Aspirin or ibuprofen lower the temperature?*

Dr. Russ Springate, Associate Professor of Family
Medicine at McMaster University:

Fever is, indeed, a mechanism the body uses to protect itself
from viruses and bacteria. Certain bugs are temperature-
sensitive and can't survive in an elevated body temperature,
so the body has a mechanism for generating that higher temperature. We don't understand the mechanism completely,
but it is part of the immune system's response to an invader.
White blood cells recognize an invader and release a natural
chemical called interleukin. The interleukin is circulated to the
brain and into the hypothalamus, which is the temperature-
control centre of the brain. We think it triggers the production
of a hormone, prostaglandin, which resets the brain's thermo-
stat. The brain realizes the body has to be warmer, and initiates
a series of mechanisms to make the body warmer, including
the common ones we know of, like shivering.

Most of the time, the temperature rises to a maximum
of about 40 degrees Celsius, which is still a safe body temperature. Unfortunately, in rare cases, the body overdoes
the job and raises the temperature too high. This can lead to

serious trouble, such as convulsions or, if it gets high enough, even death.

Let's move on to the second part of the question, and how Aspirin and ibuprofen reduce fever. Both drugs are absorbed from the stomach into the bloodstream, and circulate into the brain. Once they get there, they actually inhibit the production of prostaglandin, the hormone that signals the body to boost temperature. As a result, the body doesn't attempt to warm itself as much, and fever drops.

You might wonder if that is always a good idea. If the body has evolved this fever response as part of an immune reaction to kill bugs, should we be defeating it? This is actually a subject of considerable debate in medicine. Certainly, it's wise to lower the fever when it gets to those high ranges: 39.5–40 degrees. Again, once temperature gets that high or higher, some people are susceptible to convulsions and some children, in particular, have what we call febrile convulsions. A fever will cause them to seize, which can be dangerous. Another reason for lowering the temperature is just to feel better. If we have a simple virus, which the body is going to eliminate in three or four days, fever might be overkill. It might just be more comfortable to bring down the fever and feel a little bit better, while the immune system does its job.

SODIUM IMPLODIUM

▲

When you are stranded in a desert, I understand the necessities are
water and salt. Why, then, can't you drink saltwater when you're
stranded at sea? If it's just a matter of being too salty, then how much
seawater can you add to your freshwater to stretch your supplies?

Dr. David Behm, Associate Professor of Exercise
Physiology at Memorial University of Newfoundland:

It is certainly true that we need both salt and water in order
to survive, but unfortunately we can't use the salt in the ocean
very effectively. The first problem with seawater is that it contains a number of different salts. Normally, when we think of
salt, we think of table salt, which is sodium chloride. Saltwater
does contain sodium chloride, but it also contains a number of
other salts, such as magnesiums, sulphates, and calciums,
which we can't use.

The second problem is that when we drink water from the
ocean, the concentration of salt in seawater causes us to get
dehydrated, rather than rehydrated. In other words, we end
up losing water rather than gaining water. The first mechanism at work here is osmosis. Water always wants to move
from an area of low concentration of salt to an area of high concentration of salt. So if you're drinking saltwater from the
ocean, then the water from your body will want to move into
your intestines, in order to dilute the high concentration of salt
in your intestines. That means that water will actually be

drawn from your tissues, in order to dilute the saltwater, and you can become dehydrated.

Apart from osmosis, there is also a physiological mechanism that does the same thing. While we all need salt, too much can be harmful, so the kidneys usually trap the excess and we urinate it out. The body's response to too much salt is to urinate, and that requires water, which is going to be drawn from the body. An example of the same phenomenon is familiar to those of you who have, on occasion, gone out on a Friday night and drunk too much beer. You drink all that beer, and there is lots and lots of water in it. But you have a headache the next morning because you're dehydrated. The beer acts as a diuretic – it stimulates urine production and steals water from the rest of the body. So the end result of drinking seawater would be that you'd lose more water than you absorbed.

As to the second part of this question, you could indeed stretch your water supply a little by adding seawater. We can tolerate small concentrations of salt in water before it begins to have the effects I've described. You wouldn't be able to add much, though, because the concentration in seawater is so high – about 3.5 to 4 per cent.

Of course, the final problem with drinking seawater is that you tend to send it right back up again before it gets into your intestines. The body reacts to that much salt as if it were a poison, so keeping it down might be the biggest problem you would face.

3

Fur and Fangs:

BIOLOGY AND THE ANIMAL KINGDOM

AVIAN WEIGHTLIFTING

▲

How much of a bird of prey's weight can it actually carry off in flight?

Dr. David Bird, Professor of Wildlife Biology at
McGill University and Director of the Avian
Science and Conservation Centre in Montreal:

Some birds of prey, or raptors as they are properly named, can carry at least their own weight in their talons. For example, ospreys have been seen taking off with fish which weigh as much as they do. This means an osprey can carry at least 100 per cent of its body weight. But there are some eagles that can do better than that. There was a case of a Pallas' fish eagle, a European bird, carrying a carp that weighed much more than itself, judging by its size. The carp weighed about 5,900 grams, whereas an average fish eagle weighs about 3,700 grams. That represents more than one-and-a-half times the bird's body weight. Other eagles can also carry more than their own body weight. For example, there is a case of a bald eagle, which on average weighs about 6,300 grams, carrying a young mule deer (probably a fawn) weighing 6,800 grams. But not all eagles carry that kind of weight. Harpy eagles, which are among the most powerful eagles in the world, weigh about 9,000 grams and are known to carry sloths, which only weigh about two-thirds of their body weight.

Even the smaller birds of prey are very good at hauling weight around. The average kestrel probably weights about 130 grams, and one was seen carrying a rat which weighed about 240 grams. It may be true that smaller birds can carry a larger proportion of their weight than the large birds. Loggerhead shrikes weigh only about forty-five grams and they are capable of killing small birds that weigh about the same as themselves, such as larks. So they are probably easily capable of carrying at least 100 per cent of their own body weight.

The ability to do this comes from the power in the bird's wings. They have two sets of huge, powerful flight muscles, the pectoralis and the supracoracoideus, attached to the sternum, which work against each other and allow the bird to get off the ground. While they are very powerful muscles, the birds still need to do a bit of a hop to get going off a perch or something similar. And even these muscles won't let them get very far with a big weight. Usually these birds will just travel a few metres to a safe perch before settling down to eat what they've caught.

But many people have probably heard the following anec-dote. Back on July 25, 1977, someone described seeing not one but two giant birds swoop down on a ten-year-old boy and apparently carry this thirty-kilogram child by his shirt, a metre off the ground, over a distance of seven or eight metres. But these birds were described as overgrown vulture-like birds with white rings around their necks. The witness thought they might be condors, then, later on, decided they were immature turkey vultures. I know turkey vultures can't carry that kind of weight. They have fairly weak feet compared to most birds of prey. What they think now is that the young boy was running along with a vulture clinging to his back and at no point was he airborne.

KNOCK, KNOCK, KNOCKING
ON WOODY'S DOOR
▲

How does a woodpecker pounding on a tree keep from getting a concussion?

Dr. Alec (Sandy) Middleton, Professor in the Zoology
Department at the University of Guelph:

This question has been asked for years in ornithology. However, we do not really have a full answer. As the bill of a woodpecker slams into a tree, we would expect the braincase to decelerate rapidly, and the brain to be thrust into the front of the skull, causing concussions. But we know there is no brain damage, so the brain must be held rigidly in place, and cushioned from shock. What we do not know is the exact morphological relationship of the brain to the braincase, which is surprising, given our knowledge of bird anatomy. For example, we do not know how tightly the brain is fitted into the braincase, or if the brain is held in place by some sort of harness, like a seatbelt.

When we look at woodpecker anatomy, there are some clues we can pick up. First, woodpeckers have a flat, chisel-like bill that is straight and aligned with the floor of the skull. Most birds, by contrast, tend to have the bill aligned higher on the front of the skull. Because of this unusual alignment, the shock waves pass along the bill of the woodpecker and, instead of being transmitted to the brain, pass through to the base of the skull. It is similar to pecking the tree with your chin rather than your nose.

Second, there is the structure of the upper jaw. All birds have a kinetic hinge, which means they can elevate their upper jaw. In the case of the woodpecker, this is really pro-nounced. As well, the woodpeckers that really hammer into wood have an extension of the frontal part of their skull that comes forward over the base of the upper jaw. The combina-tion of being able to lift up the upper jaw and having a bony projection means that, upon impact of the bill with the tree, the bill is elevated until it strikes the projection. As a result, the force of impact is deflected down onto the floor of the skull. So, again, the floor takes the pounding, rather than the front of the skull and the brain behind it.

Finally, we also know that the woodpecker skull, com-pared to those of other birds, is thicker, but as with avian bone in general, is still highly pneumatized. That means there is an air space between the two layers of bone that form the skull. This structure is akin to a sandwich; there is an outer layer of bone, an inner layer of bone, and a spongy layer in between. The effect is a bit like an air bag, which can absorb the impact and cushion the brain.

FIDO'S FEELING FUNNY

▲

My dog seems to react badly to certain foods, especially chicken, as though he were allergic to them. Can animals get allergies?

Dr. Peter Foley, Resident in Small Animal Internal
Medicine at the Atlantic Veterinary College,
University of Prince Edward Island:

Allergies in animals are actually relatively common. Some dogs can be allergic to different components in their diet. It is usually a protein source or a carbohydrate source, and chicken is certainly one of many protein components that have been implicated. They can also be allergic to flea bites and flea saliva. Some animals are allergic to airborne substances, such as pollens, moulds, dust, or dander from other animals. They can also have contact allergies, where they are allergic to something they actually touch.

Human responses to allergens come in a number of forms. We might have a runny nose, swelling, itchy rashes, or difficulty breathing. But animal allergies usually manifest themselves as a skin problem with itchy skin or rashes. Sometimes that can predispose the animals to bacterial or yeast infections of their skin. It is not usually life-threatening, but it can be associated with a lot of discomfort for the dog or cat. Unfortunately, because they do not look like the kinds of allergic responses humans have, pet owners can often miss animal allergies.

It might seem odd to think of dogs developing allergies. They certainly have more of a cast-iron gut than humans do. But some dogs appear predisposed to developing allergies to substances in their foods. If you have a dog that is allergic to something in its diet, there are special foods available that are designed for allergic animals. Or you can prepare home-made food, where you eliminate the offending ingredient. If that is all the dog is allergic to, then that may very well com-pletely solve his problem.

A Foot in Cold Water

▲

During the colder months of the year, I often see gulls, ducks, and swans swimming in frigid, almost freezing water. And in winter I sometimes see birds, such as chickadees, walking around on the snow. They all do this without the aid of footwear. Without fur, feathers, or blubber on their feet and legs, how do birds keep their feet from freezing under these conditions?

Dr. David Bird, Professor of Wildlife Biology at McGill
University, and Director of the Avian Science and
Conservation Centre in Montreal:

This question has two important elements. First, the tissues
of birds' feet are obviously very tolerant of the cold – they can
get to quite low temperatures and continue to function as long
as they do not actually freeze. Second, when you think about
it, it is also interesting that the birds do not suffer from having
this large area of their bodies exposed to cold. If you were to
go out in winter with bare legs and feet you'd not only be at
risk of frostbite, but also hypothermia, since the loss of heat
through your exposed flesh would drop the temperature of
your whole body.

Birds have some interesting physiological adaptations
that allow them to tolerate cold this way. Most birds have a
complex network of blood capillaries in their feet that allows
them to reduce the circulation in their feet to a mere trickle.
Mind you, they can't drop the temperature too far, or their feet
would freeze; so they have to maintain some circulation. As a
solution to this dilemma, some birds also have a special kind
of heat-exchange system, in which warm blood going to the
feet flows past blood returning from the feet. The returning
blood is warmed by the blood going to the feet, so the heat is
conserved within the bird's body.

Feathers are also important here. You can't get a better
natural, lightweight insulation than feathers, and what birds
will often do is pull the leg or the foot up into the feathers and
sort of take turns keeping them warm. There are other tricks

they use to survive the cold in general. Birds can actually drop their body temperatures at night. For instance, a little chickadee will go into the woods. First it has to feed its face at a good feeding source, and then it will go off in a tree and simply drop its body temperature and survive the night that way.

Another thing birds can do is fluff up their feathers and then heat up the air pockets between the feathers and their body. It is just like when you go out to buy a winter coat: you do not want to buy something that is really tight. You want to buy something that is a bit loose so that you can create a layer of warm air between the coat insulation and your body.

Some birds, like grouse, ptarmigan, and even snow buntings can do something else to keep warm that is rather neat. What they do is plunge into the soft snow – they'll go right underneath the snow – and then they use their body heat to warm up a cave, like an igloo. But sometimes there will be situations where the temperature's dropped down to double digits below zero, and a rock-hard icy crust has formed on the snow and they can't get out.

A PURR-FECT PUZZLE

▲

How do cats purr?

Dr. Sherlyn Spooner, Veterinarian in Montreal with
a speciality in feline behaviour:

There have been a number of different theories about this behaviour, and the question is probably not completely answered yet. However, the most popular theory involves the cat's larynx and glottis.

The larynx, in humans, is the voice box. Cats have one as well. The glottis, which is part of the epiglottis, lies near the larynx. The epiglottis acts like a protective lid to prevent food from going down the windpipe when an animal swallows. If you were to look far down the back of a cat's throat, you will see the V-shaped opening to the larynx, which is capable of opening and closing quickly. As the cat exhales, air is being pushed out through the rapidly opening and closing larynx, past the mostly closed glottis, and produces part of the purring sound. The sound continues, but less intensely, when the cat is breathing in and has the glottis open. The sound continues even when the glottis is open due to continuous contractions of the larynx.

The question that really interests me is why cats purr. And as with most questions that have to do with the ever-elusive domain of cat behaviour, we do not have the complete picture. We know that cats purr voluntarily – they decide whether or not to purr.

Cats begin to purr at quite a young age. Kittens can start purring when they are as young as two days old. When the mother cat initiates nursing, she will begin to purr, as will the kittens when they start to nurse. Cats usually purr in response to a pleasurable stimulus such as petting or a nice scratch under the chin. There are cats that will purr in response to the presence of food. On the other hand, purring is not just associated with pleasure and comfort. Cats will purr when they are in a mildly anxious state as well. No one has done any studies to see whether there is an actual difference in the sound that is being produced in response to these two different stimuli (pleasure versus anxiety). To our ears it sounds the same, but it may be that if we were able to measure it, we would find that the two purrs are slightly different.

Purring is simply part of the cat's vocabulary, and we just do not always understand what they are trying to tell us.

AREN'T THEY TALON-TED?
▲

Do birds have a tendency to be right- or left-clawed?

Dr. David Bird, Professor of Wildlife Biology at McGill University and the Director of the Avian Science and Conservation Centre in Montreal:

This question interests me personally, because I'm left-handed. It is a question that has attracted some scientific attention because of what it tells us about the way animal brains are organized. We do not actually use the term left- or right-clawed in birds. In the scientific publications discussing animals, the term "handedness" is often used, but of course with birds we're actually talking about "footedness," since their claws are their feet and their "hands" are in their wings.

Footedness has been studied in several species of birds. One of the most interesting is the case of the crossbill. We have two species of crossbill in North America, and they have bills in which the top part of the bill crosses over the bottom. This allows them to pry and hold open the scales of cones from trees, so they can get the seeds out of them. The chances that the bird's upper bill will cross to the left or right is about fifty-fifty. The link to footedness is that the bird uses its feet as well as its beak to eat the seeds inside the cones; and the foot it uses to do this corresponds to the direction that its beak crosses. If its upper beak crosses to the right, it will be right-footed, and, in fact, that foot will actually be bigger than the left. So cross-bills are about fifty-fifty left- and right-footed.

Footedness has also been studied in parrots, because they use their feet a lot to pick up their food. Sixteen species have been looked at, and in the few studies that have been done, about three-quarters of the species bring their food to their beaks with their left foot. In another study of brown-throated parakeets, half of them were right-footed and half of them were left-footed. Pigeons have been studied as well, by sticking a little piece of tape on the end of their beak and seeing which foot they use to try to get rid of it. But they did not show any preference. Some British ornithologists did a study of pigeons, looking at them landing and taking off, and they concluded that only in landing was there even a mild degree of foot preference.

The birds that interest me the most are birds of prey, and there have been anecdotal studies that have shown that goshawks rest on the left foot more often than the right. A famous British photographer noted that quite a few different kinds of birds of prey carry their prey most often in their left foot. So I did a little study with a student in which we observed sixty American kestrels, but overall we did not find any kind of footedness in these birds.

So the best we can say in answer to this question is that it seems very clear that some birds do have footedness, but probably not all.

Stinkin' Up the Joint

▲

Last week at my school, a skunk sprayed in our schoolyard and the smell came in the windows and through the ventilator. The whole school stank for the rest of the day. We could hardly stand the awful stink. So my question is: since skunks smell so bad to us, how come skunks can stand their own smell? And do we smell bad to them?

Dr. Mark Engstrom, Associate Professor of Zoology at the University of Toronto and Senior Curator of Mammals at the Royal Ontario Museum:

As anyone who has encountered one knows, skunks do smell very strongly. When a skunk comes by your backyard, you can often tell that it has been around, even if it hasn't sprayed. We have one that comes around our yard, and every night my daughter says, "The skunk came and he left his smell behind."

The answer to how they can stand their own smell probably has to do with sensory processes in the skunk's brain. The brain does a lot of filtering of the information that is coming into it so as to make sense of it. There is so much light and sound and smell in the world being picked up by your senses that you can't possibly process it all. So basically, your nervous system ignores some of it. For example, if you're in a crowded restaurant, there are probably lots of conversations going on all around you. But humans are pretty good at shutting all that out and just concentrating on the voices they are interested in, like your companion at the table. In the case of a skunk, he's got all these smells coming in all the time, some of them very strong smells (skunks do get into the garbage, remember). I'm sure that what the skunk does is tune into the scents it is interested in, and ignore the rest – including his own powerful odour.

Even you can learn to ignore these kinds of smells. For example, I have had some fairly smelly animals in my lab. One time, someone in the Northwest Territories sent us some wolverines, which actually smell pretty bad – a lot like skunks. I had them in the lab, and I was working with them, and after a while I did not smell them any more. I thought that the smell had just faded away. At the end of the day, I got on the subway (it was about five o'clock and very crowded), but for some reason, I was given a lot of space, and pretty soon I was the only person sitting on my seat. The smell hadn't gone away – my brain was just ignoring it.

As to the second part of this question – what smells bad to skunks? – well, it is hard to know. But there is a good chance that it could be humans. We don't smell nearly as strongly as skunks, since we don't have their scent glands. But humans do have a very distinctive odour. Most other animals can tell right away that a human is in the area, and most of them will avoid us pretty well. It could be that the animals have evolved a protective disgust for our smell.

SNIFFING OUT DANGEROUS WATERS

▲

How do sharks smell blood? How much blood will attract a shark? How far away would the sharks be able to sense the blood?

Tim Low, freelance consultant in Vancouver and former Curator of Fishes at the Vancouver Aquarium:

Sharks smell by sampling the contents of the water they swim through. Water flows in through openings on the outside of the shark's nostrils. Inside the shark's head, the water passes through a funnel-shaped passage into the nasal sacs, which are lined with receptor cells. The sharks do use their other senses, like sight and hearing, to home in on the victims, but smell is their strongest tool to assist them with location. Their ability to smell is remarkably sensitive – they are normally capable of detecting dilutions down to one part blood in twenty-five million parts seawater. Under special circumstances it can be even more sensitive. We have done experiments with black-tipped sharks that had not eaten in a while and were good and hungry. They could detect concentrations of only one part blood in ten billion parts seawater. At that sensitivity, they are capable of detecting blood at quite a distance – at least a hundred metres.

PUMPING PROTEINS IN YOUR SLEEP

▲

Do animals lose muscle mass during hibernation the same way that humans do with inactivity? If they do not, why not? For example, with several weeks of inactivity, humans would have a difficult time walking. Do black bears experience the same problem in the spring?

Dr. Ken Storey, Professor of Biochemistry at
Carleton University:

The short answer is clearly no. Bats that are hanging upside down in their caves during the winter, can, after many months, still fly out the first minute of the first day after their hibernation ends. Groundhogs that have gone underground in Canada in September still have enough muscle mass to come out. They are thinner because they've lost fat, but they are capable enough to mate and reproduce almost immediately.

There are two reasons why hibernators do not lose muscle, as humans would with this kind of inactivity. The first one is simply temperature. You are at 37 degrees Celsius and your body is working very quickly, making proteins and breaking them down. If you become inactive, and your muscles aren't stimulated to build new proteins, at 37 degrees, a very warm temperature, you tend to break down the proteins that already exist very quickly. Hibernators, on the other hand, let their body temperatures drop as low as 5 degrees Celsius. They are ice-cold. If you touch a hibernating ground squirrel or bat, they feel just like a brick of ice. At this temperature the metabolism goes slower, and the animals really are in a kind of suspended animation. The proteins break down so slowly that it would take years and years of inactivity for there to be any significant loss of tissue.

The other reason is that a lot of animals are also very careful with their proteins, in a way that humans aren't. Humans have evolved in such a way that when we stop eating, and the body has to start breaking down its own tissues to survive, protein is the first thing we use. Then the carbohydrates and fats are metabolized. Animals spare proteins. They do not allow proteins to be broken down and they preferentially burn the fat that they lay down. This is just a product of our different evolutionary history.

It is interesting that the listener asked about black bears, because they are kind of an exception to the rules about hibernating animals. While other hibernators will drop their body temperatures down to 5 degrees, black bears keep their temperature around 29 or 30 degrees, and they don't shut down their metabolism that much. As a result, when scientists go into a hibernating black bear's den, often we will find them awake and fairly active. The bears manage this because they can lay down a tremendous amount of fat for the winter. During the fall, they change feeding strategy and eat very fatty foods. They lay on 20 or 30 per cent of their body weight in extra fat. Then, when they hibernate, they use fat preferentially to protein in the way that humans can't. So by ramping down their metabolism, even if it's just down to 30 degrees, and subsisting on the extra fat they have laid down, they can comfortably wait out the winter.

Another factor, which we now know from our molecular-biology studies, is that hibernators have special mechanisms in their muscles to "turn on" genes during the winter in order to rebuild the muscles while the hibernator sleeps. While the inactive muscles of humans utilize their own proteins during inactivity, hibernators, on the other hand, can turn on the genes that code for muscle proteins and, even with the inactivity of hibernation, synthesize new muscle protein. The new proteins help to bulk up the muscles so they are not atrophied upon arousal. One more interesting observation is that many of these animals are actually, even in hibernation, kind of restless sleepers, and this probably helps them maintain muscle tone. You have probably noticed that sitting still, even for an hour or so, can result in stiffness. If hibernating animals didn't move for the whole winter, they would probably be stiff, too. In fact, they move quite a lot. If you look underground at the ground squirrels, or at bats in a cave, they are wiggly and twitchy, like a small baby in a crib. These small movements keep them from stiffening up, and actually serve them very well.

AVIAN ACROBATICS

When I observe birds flying in a flock, suddenly, in the blink of an eye, the whole flock will change direction or land in a tree simultaneously. How do they do it?

Dr. Barrie Frost, Professor of Psychology and Director of the Visual and Auditory Neurosciences Laboratory at Queen's University:

It is a remarkable feat of coordination, and it certainly does look as if the birds all turn or move simultaneously. But I suspect that it is not actually simultaneous. I think it is probably more like a fast wave of activity that spreads through the flock very quickly.

It is not a phenomenon that is been well studied in birds, but there has been a lot of work done on schooling fish, which have the same behaviour. What's been shown in the fish is that they're actually turning in sequence, but the sequence is very, very fast. In one experiment, researchers shot film of fish approaching a pane of glass in a tank. The school approached and apparently all turned at once. When the researchers looked at the individual frames of film, they found that only three or four frames of film passed between the leaders encountering the wall and turning and the rest of the school turning. That is really just a fraction of a second. Birds, because they're moving so quickly, may have to react even faster than fish.

How they do this is still a bit speculative. We think they're using a remarkable adaptation that we're pretty sure evolved a very long time ago. We think it is part of a special section of the visual system in the brain, called the accessory optic system, which has to do with maintaining posture and orientation. If, for example, you take a single fish – a goldfish, perhaps – and you put it in a circular glass tank, and then slowly revolve a vertically striped pattern around the tank, the goldfish will maintain a specific orientation to the stripes. In other words, it looks like it is stuck to the stripes. Most animals will do the same thing in order to stabilize themselves to cues in the world. This is known as an optimotor response.

With schooling fish, and possibly birds in a flock, this special part of the visual system works so that the animals orient themselves to each other, and so the same system helps them stay in formation. Therefore, if a bird in the middle of a flock stabilizes himself in relation to his neighbours with this opti-motor response, it is going to turn when the flock turns.

TIME TO FLY
▲

How do migrating birds who go south for the winter know when it is time to fly back north?

Dr. John Ryder, Professor Emeritus of Biology at
Lakehead University:

This is a great question, and I have thought about it for a long time. After all, there are clear cues for when birds should head south – changing temperature, trees dropping their leaves. But it's less obvious what would tell them to leave Florida, for example, which has a more consistent climate.

In examining this problem, we ornithologists look to the environment for something that will be a constant cue every year. The only answer we can come up with is the difference in the amount of light every spring versus every fall. The light increases in the springtime and decreases in the fall. We assume that birds cue into this difference to head back north.

We know, for example, that the increasing length of the day in the spring has an effect on birds' reproductive systems, helping them determine when to nest and lay eggs. So it probably can also trigger a kind of restless feeling that culminates in spring migration.

Birds have a little gland in the mid-brain called the pituitary gland, which is very sensitive to changes in light. Once the increase in light occurs in the springtime, the pituitary gland starts to secrete hormones that affect the reproductive system. That is the immediate change that occurs. But birds in the springtime on their wintering grounds start to behave differently, too. They go through a process of migratory restlessness, and when the light increases to a certain point, they will just take off and fly north.

Of course, you have to wonder why birds migrate at all from their comfortable and warm wintering grounds. We think it just boils down to evolutionary experience. Birds have been around for sixty-five million years or so, and they have adapted to maximize their reproductive potential in the environment. While the wintering grounds have good weather, there is a lot of competition there for nesting areas and food. Up north, however, there is lots of nesting habitat and food, so they can nest, feed, and reproduce. When things get cold and nasty, they can head down south again. Basically, there are resources up north to be exploited, so birds developed a way to take advantage of that. A bird's evolutionary drive is just to make more birds, and this is a way for that to happen.

FLUFFY'S FABULOUS
FIDDLEHEAD-FREE DIET

▲

How can cats stay healthy if they don't eat vegetables?

Dr. Susan Little, Veterinarian and owner of two
feline specialty practices in Ottawa:

The answer to this question lies in the fact that cats, unlike
people, are true carnivores. Their evolution has adapted them
to satisfying all their nutritional needs from animal sources,
and so they have no need and little inclination to add vegetables to their diet.

In part, they have different nutritional needs than we do.
The amino acid taurine, for example, is one of the building
blocks for our body tissues. Humans can synthesize it within
our bodies, but cats must get it from their diet, and it is not
present in vegetable sources at all – only in animal sources. A
deficiency of taurine in a cat's diet will lead to serious eye and
heart problems. On the other hand, we need to eat fruits and
vegetables for the vitamin C they contain, and will become
very sick if we don't get it. Cats and dogs can make vitamin C
themselves from other constituents in their diet, so they can
do without "an apple a day."

Cats also cannot use the beta-carotene in foods like
carrots, which we would use to make vitamin A. They still
have the need for vitamin A, but they must get it in their diet
on a preformed basis, and that is only found in animal tissues,
primarily in the liver of the animals that they would eat. That

is also why domestic cats and cats in the wild have such a taste for the parts of the animal we often aren't fond of. They will eat all the innards except for the intestine. They will often eat the head of their prey first. This is probably because those parts supply important nutrients for them that we get from other foods.

Of course, the commercial pet foods that are prepared for cats are quite different from wild mouse heads, but we have learned over the years what sort of supplements we need to add to these foods. We have to make sure there is enough vitamin A and taurine, plus four times more niacin than for dogs, for example. Even so, the commercial foods are a bit different from a natural diet. There is about two to four times more fibre in prepared cat food than in a wild diet. It is quite a bit less expensive to use non-animal sources for cat food, and so most commercial foods have vegetable matter in them, but it is supplemented to make it nutritionally complete.

HAPPY TO SEE ME?

▲

Why do dogs wag their tails? What possible evolutionary benefit could be derived from doing this?

Dr. Norma Guy, Assistant Professor in the Department of Anatomy and Physiology at the Atlantic Veterinary College, University of Prince Edward Island:

There is an evolutionary reason why dogs wag their tails, and it has to do with where dogs come from. Dogs are descended from wolves, and wolves live in a complex social group. Within that group, they need to have a clearly defined set of social behaviours to allow them to communicate their attitudes and intentions in order to prevent fights and maintain the social organization. Tail-wagging is just one of the behaviours that wolves demonstrate in order to maintain social harmony in their group. They have a number of other things they do, including postures such as head position, ear position, and facial expression.

Tail-wagging is a quite important behaviour, because wolves use it to defuse tense situations. When a subordinate wolf greets a more dominant wolf, for example, and wants to indicate that it doesn't mean any harm, it usually approaches with a lowered body posture and a very freely wagging tail. That indicates to the more dominant wolf that this is just a little visitor that isn't posing any threat or challenge. The dominant wolf typically doesn't wag in this situation, but adopts a more upright stance to acknowledge the behaviour of the subordinate animal.

When it comes to dogs, tail-wagging is something they do pretty indiscriminately whenever they are happy or excited. That is probably because it is a trait we may have inadvertently selected for in the course of the millennia that we have been breeding dogs. The fact that different breeds do it a little more than others is an indication that this might be the case. It also makes sense that we might select for this behaviour since, after all, we like it. It is easy to understand. It is a very overt indication of their mood, and it is a nice greeting behaviour that people tend to respond to. Dogs are also probably signalling to us the same thing that wolves do when they wag their tail: "I'm happy to see you, I mean you no harm, let's talk." It is probably not so much that they see us as dominant precisely, but it is an acknowledgement of our place in the family.

All tail-wagging isn't quite alike, though, and people should be aware of this. That is why we tell children not to believe that every dog with a wagging tail is a friendly dog. Tail movements can have different meanings. If we look at the classic sort of tail wag, it is a low, broad wag at a medium-fast speed, and that usually means that the dog is friendly, and that is fine. When you see a dog with a tail tucked well down between its hind legs and just the tip of the tail is wagging, and the dog is kind of crouching, that might be an indication of fear or anxiety in the dog. It is probably best not to approach because the dog might be easily spooked and react unpredictably.

Dogs that are showing aggression may hold their tails very upright, and move their tails in very stiff, short sort of wags. It is a good idea to be careful of those animals as well. Generally, though, if you are seeing that big exuberant wag back and forth when you get home from work or come back into the house, it means, "Hi, I'm happy to see you, you're the boss," and that's a pleasant greeting.

FELINE FIXATION

▲

Why does my cat appear to stare at nothing? Does he see things I can't see? Do cats see phantom-like energy, or is my cat just messing with me?

Dr. Donald Mitchell, Professor of Psychology at Dalhousie University, and a leading authority on cat vision:

I think a lot of cat owners will have observed something like this, and I suspect what is happening is that the cat is reacting to something it saw move, something it could not identify or locate accurately. The cat might have seen a shadow or a real object moving, and that alerts it, and then it keeps staring at that spot, waiting for it to move again. After a time, if it does see something move, then it will move towards it if it is interested, or away from it if it feels threatened. We just see the cat staring at nothing, and assume that it is behaving strangely.

It is probably a kind of hunting behaviour. Cats, through long experience, know that animals that are potential prey – mice, for example – will often freeze when they see a predator like the cat. The mice know that as soon as they move, the cat will pounce, so they stay frozen. In a sense, it's a game played by both participants. The cat knows that it can't just glance over to where it saw some motion and then satisfy itself very quickly that there's nothing there. It has to wait, sometimes for a long time, because the potential prey is still and possibly well concealed.

Interestingly (and this may not be germane to the listener's question), I have only found two instances where cats are superior to humans in vision. In most other instances, they are inferior. One of the instances where they clearly are superior is due to their absolute sensitivity to light. They can detect lower light levels than we can – the reason for their better ability to see in the dark. They are probably two to three times better at that than humans. That probably can be attributed to the fact that they have a reflecting layer at the back of their retina called the tapetum, which, in a sense, doubles the opportunity for a rod photoreceptor to detect light.

The second area in which they out-perform humans is in terms of seeing large objects, of a certain size, at lower light contrast levels. A cat can see objects subtending a visual angle of about two degrees at a lower contrast than a human can.

NAVEL-GAZING
▲

Do animals have belly buttons and what happens to them after birth?

Dr. Nora Lewis, Associate Professor and Veterinarian
in the Department of Animal Science at the
University of Manitoba:

Yes, indeed, some animals do have belly buttons. To be specific, mammals have belly buttons, but we usually can't see them. The reason only mammals have belly buttons is that only mammals have a specialized organ called the placenta. The placenta supplies nourishment for the fetal mammals in the womb and is connected to the fetus by a cord called the umbilicus. After birth, the umbilicus is separated, and the end attached to the baby dries up and drops off, leaving the belly button as the only sign of the connection.

We don't normally notice the belly button on other mammals simply because they're usually covered with fur. But if you shave a mammal the belly button is quite apparent. If you have ever had a pet cat spayed, you may have noticed the cat's belly button on its shaved abdomen. It is in roughly the same position on the abdomen as a human belly button. Animal belly buttons don't usually look like ours – they don't have "innies" and "outies." They are very flat, and usually all you can see is a little white scar.

Interestingly, some non-mammals have something like a belly button. We recently had four baby snakes born from a species where the infants are born live – not hatched from eggs. These live-born snakes actually had a connection to their abdomens that looked like an umbilical cord, and after it dropped off, it left a mark that looked a lot like a belly button. But it soon disappeared, and since it wasn't really the scar from a connection to a placenta, you would have to say it wasn't really a belly button.

4

Creepers and Crawlers:

THE INSECT WORLD

Saving at the Sperm Bank

▲

I understand the queen bee mates once and for four to five years afterwards produces offspring. What preserves the sperm or fertilized embryos?

Dr. Mark Winston, Professor of Biological Sciences at
Simon Fraser University and a beekeeper:

The queen bee has a special structure near her ovaries called a spermatheca, a sac that can store sperm almost indefinitely. It is a small, soccer-ball-shaped sac, fairly thick in consistency and hollow inside. Attached to it is a gland that secretes nutrients for the sperm. With those nutrients, the queen bee can keep the sperm alive for years – as many as eight years in some cases.

The queen is only keeping sperm alive; she's not preserving embryos. When the queen lays eggs, they are released from her ovaries one by one. At the same time, she releases a single sperm to fertilize each egg. The sperm and egg come from separate places, and meet at the tip of the spermathecal gland. Then they are deposited into the comb. She's doing her own fertilization in her body. Once the queen mates, drone bees have nothing further to do with fertilizing her eggs.

This strategy probably evolved because mating, for a queen bee, is quite a bit more dangerous than mating is for us. The queen bee goes out and flies around, where she could be the victim of predators. She could also get blown away by wind, or just get lost. As the colony is going to invest a lot of energy in its queen, the worker bees want to know she's going to be safe, and that she's well mated. So she goes out at the very beginning of her life, takes two, three, or maybe four flights, and at the end of that time, which may only take a day or two, or even just an afternoon, the workers recognize that she's mated. Now they know they can invest energy into building combs and rearing brood and all the sorts of things they need to do to keep the hive alive. They want to keep her safe and not let her go outside again.

Bees aren't the only insects with a spermatheca. Any insect with a long-living queen will have the same kind of structure and will have glands that provide nutrients for the sperm. Many insects do live for many years, and the way they mate and keep the sperm alive is through this specialized structure. It is common in social ants, some of which live up to twenty years.

A POETIC WORM

▲

An old prospector's song ends each verse with the line: "'til the ice worms nest again." Are they fantasy or do ice worms really exist?

Dr. Darryl Gwynne, Biologist in the Department of
Zoology at the University of Toronto at Mississauga:

Yes, they really do exist. Most people know them from Robert Service's poem "The Ballad of the Ice-Worm Cocktail," in which a fellow comes into a bar in Dawson, Yukon, and is served a cocktail with ice worms in it. Someone in the poem asks the question, "what's an ice worm?" and is told they live on the Mountain of Blue Snow. The poem also claims they survive by eating each other's tails. They are certainly part of Canadian literature. Of course, in the poem, the worm turns out to be just a "stick of stained spaghetti." But Robert Service had probably heard of the real ice worms that live on glaciers.

Their scientific name is *Mesenchytraeus solifugus*, which means, "sun-avoiding worm." They are members of the true worms, the group that includes the earthworm. In fact, there are close relatives of worms that live in the tundra soil, so maybe that is how the ice worms got onto glaciers in the first place.

There is not much to eat on glaciers, so apparently ice worms live on pollen grains and algae and anything else that can blow onto or grow on the surface of the ice in the summer. These worms are quite temperature-intolerant. They do not like warmth. In fact, they'll die when they are warmed too much, and this happens if it gets much above –5 or –6 degrees Celsius. The narrow temperature range of the worms may explain why they've been found several feet deep down in the cracks of glaciers.

Ice worms are pretty nondescript-looking things. Although the one in the Robert Service poem is described as "four inches from its tail-tip to its snout," they are actually only about an inch long (or twenty-five millimetres) and they are coloured a darkish red, or sometimes black. They can sometimes be found in very large numbers, with colonies containing hundreds of worms per square metre. This is when they're most likely to be seen, in the spring when they reproduce on the glacier.

STICKY FEET

▲

Why don't flies fall off the ceiling?

Dr. Hugh Danks, Entomologist with the Canadian
Museum of Nature in Ottawa:

The first thing to note is that flies are light in weight and they're small in size. They have an external skeleton that is relatively light, and, in contrast, humans have a heavy internal skeleton made of bones. If we were to try to hang from the ceiling, we would need some pretty sophisticated hardware. But flies are light, so they don't need the hardware. They are light enough that they aren't really fighting much gravity to stay suspended.

Flies use tiny pads on their feet to hang from the ceiling. Each foot has a couple of claws, mainly used for hanging on to rough surfaces, and two tiny pads that allow them to attach themselves to smooth surfaces. We are not certain how these pads work, but there are very minute hairs on the pads and an oily secretion. It seems that, when the pad is applied to a surface, molecular forces, in effect, stick the fly to the surface and keep it there. There are really micro suction pads on the bottom of their feet.

Staying on the ceiling is a balancing act between the pads being so sticky that the fly can't release from the surface, and not sticky enough to hold the fly up. But luckily for the fly, it is light enough not to need a very strong glue, and, along with the action of the pads, its walking and flying strength can break the bonds with the ceiling when it wants to move its feet. The system works well enough that many insects use it. It is a pretty common strategy.

As well as walking on the ceiling, a fly has to get there in the first place. How it does that is an interesting question that was solved a number of years ago using high-speed photography. Flying upside down is a bit of a trick. You might think that flies would do a barrel roll like an airplane, so they would first turn upside down and then land on the ceiling. But if they're upside down, both their wing action and gravity are pulling them down, so flies have developed a different strategy. As a fly gets close to the ceiling, it flies in at an upward angle, and then touches its front feet to the surface. When the front feet touch down, the fly's momentum pivots it over its front feet and it flips over to land upside down. It is like a trapeze artist with his hands on the bar, whose feet flip up at the highest point in the arc.

DROWSY DRONES

▲

Do insects sleep?

> Dr. Hugh Danks, Entomologist from the Canadian
> Museum of Nature in Ottawa:

This is a surprisingly complex question, because sleep, as we usually understand it, applies just to humans and other mammals, when their eyes are shut and they are unconscious, and so on. But insects can't shut their multi-faceted eyes. And while we can control when we're active and when we're resting or sleeping, insects are cold-blooded, so they are very much at the mercy of the environment when it comes to their activity. When it is cold, they become inactive. Insects also show patterns of activity related to the time of day. Some are active in daytime and inactive when it is dark, and vice versa. But when insects are inactive it is not really sleep as we understand it – it's more like rest.

Insects rest on a daily basis, but also for much longer periods during difficult seasons like the winter. One kind of rest during these periods is called quiescence, when it is simply too cold for them to function. Another kind of insect rest is called diapause. That's when insects use cues from the environment to halt their life cycle temporarily. When the days gets shorter, indicating that the winter is approaching, insects go through a kind of shutdown of development and drastically reduce their metabolic rate. In this state they can stay dormant for very long periods. There are some species that can be in diapause for many years at a time. These insects can be active for just a few weeks or months and then go dormant for more than ten years.

Quiescence is a direct response to adverse conditions. For example, development stops when it is too cold because metabolism cannot go on at low temperatures. Diapause, on the other hand, is a programmed response to adverse conditions. For example, the shortening day length indicates winter is coming, and the insect prepares in advance for the onset of cold weather. It is a kind of early warning system.

ARACHNID ANTICS

▲

Why don't spiders get stuck on their own webs?

> Brad Hubley, Entomology Collection Manager at the
> Royal Ontario Museum:

There are actually two parts to the answer to this question. The first is that only part of the spider's web is sticky. The spider makes two types of thread, and one has glue on it and the other doesn't. You're probably familiar with the webs of a common garden spider, which is a yellow and black spider that spins what we think of as the typical web. It looks kind of like a bicycle wheel with long spoke-like threads radiating outwards, and circular threads woven around the spokes. These are called orb webs, and the important thing to know about them is that only the circular threads are sticky – they're called the catching spiral. The spokes radiating outward aren't sticky, and those are the threads the spider walks on as it traverses its web.

In addition, the spider has specially adapted claws on the tips of its feet, so that when it does need to move on one of the sticky threads, it wraps the claw around it, keeping the contact to a minimum. The spider also knows that when it does touch a sticky thread, it isn't wise to thrash around. When a grasshopper or a fly is caught in the web, it tends to thrash about to try to get free, which just tangles it up further. When the spider gets stuck on a thread, it calmly recognizes that this is just its own web, pulls itself gently free, and continues about its business.

Of course, orb webs aren't the only webs spiders spin. The cobwebs you find in your basement or attic are spun differently. They are a mass of fibres oriented in all directions, but cobwebs aren't necessarily sticky. They catch insects simply by entangling them in many strong threads of silk. Then the spider comes out, bites its prey gently or quickly to paralyse it, and backs off to see if it starts to slow down. If it does, the spider comes back in and bites it again, and wraps it up. The spider avoids getting caught by carefully and meticulously navigating the many tendrils of the web.

DROSOPHILA DERIVATION

▲

*Where do fruit flies come from? If I bring a banana into what appears
to be a fruit fly–free room, and it is left there unpeeled, those magical
little critters appear out of nowhere.*

Dr. Paul Lasko, Professor and Chair of the Department of
Biology at McGill University:

Before I give you the answer, you are going to have to let
me hold forth on the wonders of the fruit fly, since they are
the animal I study, and they really don't get enough respect.
For most of us, *Drosophila melanogaster*, which is their Latin
name, are a bit of a pest, buzzing around the fruit bowl, but for
scientists, they are really wonderful creatures. There are lots
of different species, especially in Hawaii, where there are some
that reach an inch long and are quite beautiful, with lots of
exotic colours. Of course, from a scientific perspective, they
are wonderful animals to study because they are so safe and
easy to work with. They are especially important for genetic
work, and much of what we know now about genetics was
discovered by people working with fruit flies.

I have some sympathy with the listener, however, as it can
be a bit disconcerting to come downstairs one day and find a
flock of fruit flies buzzing around the fruit bowl, apparently
having appeared out of nowhere.

They didn't appear from nowhere, though. The fruit flies
were there all along, but they were just in a different form. As
most listeners will know, fruit flies hatch from eggs. What

doubtless happened was the bananas you brought home from the store had a few fruit fly eggs on them and, over a few days, these eggs developed into larvae and then into fruit flies. Fruit fly eggs are very small, and very hard to see. They are less than a millimetre long, and they tend to be a kind of cloudy white, which is quite hard to spot on a piece of fruit. The eggs were probably laid in a small bruise on the banana, and the bananas were probably close to ripe when they were purchased. Fruit flies enjoy fruit the same way we do – ripe or even close to fermenting. Originally, when people grew fruit flies in laboratories, they used banana mash as a culture to grow them on.

Interestingly, your fruit flies may have come from quite a long distance away. It takes about two weeks for a fruit fly egg to develop into a fly, so it is entirely possible that the eggs were laid before the fruit was picked, and enjoyed a several-thousand-kilometre trip to your neighbourhood grocery before you took them home.

Remember that fruit flies are completely harmless. They are not the kind of insects that bite, but if you are really determined to do something about them, there are a few solutions. You could very carefully scan your fruit before you buy it; but remember the eggs are very hard to see. You can also store your fruit in the refrigerator instead of on the counter top. While the eggs can tolerate cold, when they develop into larvae, the cold will kill them. If you already have fruit flies buzzing around the house and want to get rid of them, the simplest thing to do is to remove the fruit, and they will either go elsewhere looking for something else to eat, or just die. If you take the fruit away, the flies don't last very long at all, probably about three days. You could also try to catch them.

Try leaving a cup around with a little bit of fruit juice in the bottom. If you have some yeast, you can sprinkle yeast into the fruit juice. They love that. They also rather like beer. The flies will fly in there to try to drink the juice or beer, and they will get stuck. Eventually you'll find them floating in the liquid.

However, you should look at your fruit flies with a little bit of respect before you trap them in the fruit juice, because they've been the topic of a great deal of scientific work over the last eighty or ninety years or so.

5

Solids and Surfaces:

MATERIALS SCIENCE

Egg-splosion

▲

My sister-in-law took a partially peeled, partially cooked egg and put it into the microwave to finish cooking it. Then she put it under running water, finished peeling the egg, took it out from the running water, and it blew up. Why?

Dr. Mary Anne White, Killam Research Professor in
Materials Science in the Department of Chemistry at
Dalhousie University:

The egg exploded because there was a compelling force. That force would have been the buildup of pressure inside the egg. As you start to heat the egg, gases are given off and, because the temperature is going up, the pressure of these gases can get quite high. If they aren't released, they will build up and up and you really have a time bomb just waiting to go off.

While the listener thought the egg might have cooled off after it was taken out of the microwave, that wasn't long enough to allow the gases to cool. So the pressure of the little bomb was still there. But the most important factor is the structure of the egg itself. Just inside the shell of the egg is a tough membrane. This surrounds the whole egg and is very difficult to break. You can see it when you peel a hard-boiled egg – it is the skin just beneath the shell. This membrane was holding in the pressure from the gas created during the heating.

But this membrane isn't perfect, and some of the spots will be thinner than others. All that's needed is for one of these spots to rupture and the egg will explode. That's probably what happened when she finished peeling the egg. It is just like putting a small hole into a balloon, causing all the air to rush out from that one point and break the balloon. Cooking eggs in the microwave needs some special precautions. The most important advice is to poke them carefully. You wouldn't want to cook the egg in the shell in the microwave, because that would really hold in the pressure with the membrane. But even when it is out of the shell, you need to poke the second membrane that's around the yolk in an egg. Otherwise, you'll end up with yolk and everything else all over your microwave.

TREADING ON DANGEROUS GROUND

▲

What is quicksand and could you die in it?

Dr. Darrel Long, Professor of Earth Sciences at Laurentian
University:

Quicksand is really a mixture of sand and water, or sand and
air, which looks solid, but "fails" when it is disturbed. What
you really have is very loosely packed sand. In normal sand,
the grains are all tightly packed together forming a solid mass.
But sand grains aren't all round. Many of them are elongated,
so when they are clustered together there can be a lot of spaces
between them. These spaces are filled either with air or water.
When the ground is disturbed, they collapse and there is an
upward movement of the water or air that's in them, and this
causes the ground to collapse.

The most common types of quicksand form when there is
water coming up underneath the sand. The main places you'll
find this are beaches or on the downward side of large sand
dunes. Another place you'll find them is where water comes
up in the form of springs, such as those that develop at the
base of alluvial fans. The water moves up through the sediment and produces a pipe-like structure, and on the surface
there is a little cone-shaped volcano. They are very active,
appearing to boil on the surface because there is so much sand
in them. They are really beautiful structures.

But if you were to step on one, you'd go down very quickly.
From my experience, pulling a canoe across the front of a
delta, you'll sink quickly because, essentially, there is a fabric
collapse and an instantaneous release of water or air. The
ground seems to turn to liquid because there is no adhesion
between the grains. Friction is lost as the water moves up

through the sediment. You'll sink down maybe a metre and then stop.

There are lots of myths from the movies about people sinking slowly into quicksand and drowning, but these don't reflect reality. You're unlikely to sink much more than a metre because, by then, the density of the compacted sand material is generally higher than your density. So, as long as you don't panic, you shouldn't die. You should just be able to float or walk to safety.

PERFORMING UNDER PRESSURE

▲

Why are tennis balls vacuum-packed, and why do they lose their bounce quickly once you start hitting them?

Dr. Barbara Frisken, Associate Professor in the Physics
Department at Simon Fraser University:

To begin with, let's correct a misconception, which should give some hints to the answer to the listener's question. Tennis balls are not packed in a vacuum. In fact, it is just the opposite: they are packed under pressure. That hissing sound you hear when you open a can of balls isn't air getting in, it is air escaping. Tennis balls are inflated, like most other balls. But unlike basketballs and footballs, they don't have a valve for putting in new air. So after they are made, they are packed under pressure to make sure they don't lose any air while they are on the shelf. That way, they will have a good bounce when they are opened for use on the court.

That gives a hint to the answer to the second part of the question. Tennis balls, like other inflated balls, start to lose air as soon as they are released from the high pressure of the can. Molecules of air slowly migrate out through the rubber shell that is underneath the fabric cover of the ball. Interestingly, tennis balls are manufactured with different pressures for different altitudes, so that they will all have the same "feel," whether you are playing at sea level in Halifax, or in some high-altitude Rocky Mountain resort.

You can try to offset the pressure loss in the balls by storing them under pressure. I found information on the Web about at least one contraption that will do this. You put your tennis balls in an airtight container, seal it, and pump in extra air. The tennis balls will be sitting in higher pressure and they are less likely to lose their internal air. I gather this is not considered acceptable for experts and the professionals, but it is probably good enough for everyday players.

PROBING THE PITA POCKET PROBLEM
▲

Why is it that pita bread has a pocket?

Dr. Harry Sapirstein, Associate Professor in the Department of Food Science at the University of Manitoba:

Another way to put this question is: why doesn't normal bread have a pocket? The difference in the way you bake the two breads explains both questions. In normal bread-making, what you have is a twenty- to twenty-five-minute baking process at relatively low temperatures of about 220 degrees Celsius, which is around 405 degrees Fahrenheit. What this means is that, with the long baking time and low temperature, you have a slow heat penetration.

With pita bread, in contrast, what you have is a very thin sheet of dough which has been essentially de-gassed because of the way it has been rolled into sheets. That piece of dough, in both commercial production and in domestic baking in the Middle East and North Africa where pita bread originally comes from, is cooked on a heated surface at very high temperatures, maybe double the temperatures that one would use in the oven for conventional bread. It is also cooked very briefly, for perhaps just half a minute or so per side. In fact, you can make pretty decent pita bread in the oven, as I have done in the past. All you need to have is a flat sheet of dough, and an oven at maximum temperature. The baking time for that three-millimetre- or four-millimetre-thin pancake of dough is about sixty seconds to maybe a minute and a half. It is really very quick.

So the shape of the dough and the speed of the baking is what gives you the pocket in the middle of the pita. Part of the secret is ethanol, the same kind of alcohol you find in beer and spirits, which is actually produced by the yeast in the bread. The other part of the secret is the water in the dough. At the high temperature of the pan or oven, the water and ethanol evaporate very quickly into steam. That water is contained in the bread dough in sealed cells. As it evaporates, though, it ruptures the protein membranes around the cells. So looking at the bread as a whole, what you have is an almost catastrophic rupturing of gas cells happening over a matter of seconds, which creates essentially one continuous, very large cell in the pita bread.

You might think that the bread would just blow itself apart under these circumstances, as that last big cell ruptures. But what seems to happen is that the crust that forms on the outside of the dough forms a barrier that contains the increasing pressure of the gas in the dough. The crust is stronger than the inside of the dough, and so it doesn't break open. Looking from the outside, what you will see is the pita inflating like a balloon. Once you take it out of the oven, the gas slowly escapes through the crust, which isn't airtight, and the pita deflates. But the deflated bread still has the space in the middle – the pocket in the pita – which allows it to make that ideal dripless sandwich.

6

Puzzles and Problems:
THE WORLD OF PHYSICS

GETTING AHEAD OF YOURSELF

▲

I know several people who've had the misfortune to lose a wheel from a car or trailer and have watched the wheel roll away ahead of the vehicle. Why does that happen?

> Dr. Ron Lees, Professor Emeritus of Physics at the
> University of New Brunswick:

This question strikes close to home because my parents had a wheel come off in the wilds of Wyoming once, and it rolled ahead of them down the road.

There are two possible things going on when this happens, depending on whether the wheel affected was attached to the drive shaft or not. You have to look at the forces acting on the wheel and on the tire itself.

Looking first at the wheels that are attached to the drive

shaft, these turn because they're attached to the engine of the car. But the axles are resting on the tires and the tires are resting on the ground. So they're supporting the weight of the car. When the axle starts to turn, as the car is accelerating, this distorts the tire a little, like winding up an elastic band. This produces a force forward. If the wheel comes off, this force is acting on the tire alone, which has much less mass than the car. So if the wheel falls off, it will jump forward because, at that instant, the force is the same, but the mass is much less. As good old Isaac Newton told us, acceleration is force divided by mass. Make the mass smaller and the acceleration is going to be a lot bigger.

But not all the wheels are attached to the drive shaft. When it is an unpowered wheel, a second phenomenon takes over. Like the drive wheels, the other wheels are also holding up the weight of a vehicle. When you look at the wheel on a trailer, for instance, you'll see it is flattened at the bottom. When the weight of the trailer is removed, the wheel is no longer flat, and it actually expands. Its radius gets bigger and it rolls faster than when it was on the trailer. It is still spinning at almost the same rate, but the bigger radius means it is going to speed along faster, and overtake your car.

GOING UP?

▲

If you plant a bulb incorrectly, it always grows the right way up. Why?

Dr. Andrew Riseman, Assistant Professor of Plant
Breeding in the Faculty of Agricultural Sciences at
the University of British Columbia:

The key to the plant's ability to do this is that it can sense
gravity, an ability we call geotropism. Within some cells of the
plant's tissues are structures that are basically little granules
of starch. These granules settle on the bottom of these cells,
and become the reference point for "down" within the plant.
The plant "senses" the presence of these granules and a
cascade of biochemical signals and hormones causes the cells
to divide in an appropriate direction, which leads to growth
in that direction. In the roots, the response is usually to trigger
growth downward, and in the stem of the plant, the response
is upwards.

One interesting thing is that these granules are concen-
trated in specific cells in the tissues. We know, for example, that
if you take a laser and blast out certain specific cells in a root
tip, then the roots don't know which way to grow any more.

Gravity is just one of the signals that a plant can respond
to. Most gardeners will be familiar with phototropism, the
plant's ability to grow towards a source of light. The plant can
also respond to the presence of water or nutrient pockets in
the soil, so that roots will grow towards these. Plants are pretty
flexible this way, so you needn't be too concerned about plant-
ing your seeds or bulbs right way up. They will find their way.

MAY THE FORCE BE WITH YOU

▲

I have heard that centrifugal force doesn't really exist, that it is an imaginary force we talk about to explain the reaction we feel to the real force involved, centripetal force. Could you elaborate on this?

Dr. Jasper McKee, Professor Emeritus of Physics at
the University of Manitoba:

I don't believe that the centrifugal force is real. It's a fictional force that arises from the inertia of an object. However, the idea can be useful in trying to interpret the behaviour of bodies in what are called non-inertial or rotating systems, which aren't adequately covered by Newton's Laws of Motion. Imagine a rock on the end of a string being whirled around in a circle. It certainly feels like something is pulling the rock outward, but this is evidence for a centripetal, not a centrifugal, force.

In the absence of the string, the rock would be moving in a straight line, according to Newton's First Law of Motion. So the string is constraining it from doing that. Subjectively, you have the impression of an outward pull, but really the rock is being pulled back in. Centrifugal means acting outwards, but the force on the rock is actually inward, hence, centripetal. In fact, the rock wants to fly way. If the centrifugal force were a real force and I cut the string, the rock would be expected to move radially (directly away from the centre) outward and disappear. Of course, it doesn't do that. Anybody with experience will know that. If you cut the string, it moves tangentially (at right angles) to the circle orbit and goes off sideways. That is how a slingshot works. This shows that when you cut the string, inertia takes over and the centripetal force disappears. Newton's First Law takes over and the rock continues, in uniform motion, in a straight line until something happens to it.

But even though the centrifugal force is a fiction, it is a convenient one for describing some human experiences. If you're in a rotating system, what you feel appears to be an outward-pushing force, rather than a centripetal force.

Imagine yourself as a passenger in a car that is going around a tight left-hand turn. As the car turns to the left, the tires interacting with the pavement generate a centripetal force and the car turns left. Inside, you have inertia and you are moving in a straight line. By the law of inertia, you will continue in a straight line, despite the car's motion to the left. So you find yourself pushed against the right passenger door as this occurs, and you think, goodness, there is a force trying to pull me out. But the reality, if you look at the system from the outside, is just like the rock on the string: a centripetal force is acting. So there is perception and reality. We feel a centrifugal force, but the fundamental physics is the centripetal force.

MIXING IT UP
▲

The gases in the atmosphere have different molecular weights. Why don't they separate out and form layers?

Dr. Moire Wadleigh, Associate Professor in the
Department of Earth Sciences at Memorial
University of Newfoundland:

In fact, we do have that kind of layering. The whole of the Earth is layered. We have denser rocks in the interior and less dense rocks at the surface, and that layering continues out into the atmosphere. If we look at the whole atmosphere, which extends out about four thousand kilometres, we find it consists of a series of concentric shells of gas, each with a different chemical composition. But the shell closest to the Earth, the one we live in, is homogeneous in its composition, largely because of the effects of weather patterns.

This layer is mostly made of nitrogen, oxygen, argon, and carbon dioxide. The influence of temperature (the variation in temperature both vertically in the atmosphere and from the Equator to the poles), sets up a lot of instability in the lower part of the atmosphere. The sun's energy comes through the atmosphere and heats the surface of the Earth. This heat is then transmitted to the lower part of the atmosphere, making it warmer at the bottom than at the top, and that makes it thermally unstable. So, by convection, the atmosphere tries to redistribute that heat, to move it from the Equator, where it is warmer, to the poles. These convective circulations mix up the lowest part of the atmosphere. These same phenomena are responsible for giving us our weather patterns. These weather patterns are a little like stir sticks in our atmosphere.

However, even if the impossible were to happen and weather systems were completely shut off around the world, the gases of the lower atmosphere still wouldn't stratify by density. This is because thermal oscillations, which affect all molecules at temperatures above absolute zero, cause the gases to diffuse and mingle with each other. Diffusion is so much slower than the stirring produced by weather that it isn't really a factor in practical terms. For example, during temperature inversions over large cities, diffusion can't act fast enough to prevent a buildup of pollutants. We have to wait for a stiff wind to stir up the air masses and blow the pollutants away.

This is what's happening close to the surface; but higher up, away from the region influenced by weather, we do start to see layering. About eighty kilometres up, the layering starts by chemical composition, according to molecular weight. So the first layer we see is molecular nitrogen. And then we get a layer of atomic oxygen, which are single oxygen atoms. After that, we get into helium and then hydrogen and then we're pretty much into space.

FISH FRY
▲

What happens when lightning strikes a lake or an ocean? Does the bolt continue to the bottom or is the energy dissipated in the water? What would happen if you were nearby in a boat? And do any fish get electrocuted?

Dr. Peter Watson, Professor of Physics and
Dean of Science at Carleton University:

This turns out to be a very interesting question. The standard picture of a thunderstorm begins with a very big cumulonimbus cloud. There is a very large negative charge on the bottom of the cloud, and this produces a positive charge on the Earth. A lightning strike works by a header going down from the cloud, with a return stroke coming back up from the land. The same thing would happen on the surface of the lake. It would be a very large charge right on the surface, because a lake's not a good conductor. There would be no electric current in the body of the water at all, but a current on the surface. It would not extend down more than a few millimetres, but it would spread out about five or ten metres from the actual strike.

If you were on the lake when lightning struck close by, you would probably be fairly safe. Almost certainly, if you were in a metal boat, the current would just flow around you. Even if you were in a fibreglass canoe, you would probably be well enough insulated. So, unless you actually get hit, which is somewhat discouraging, there would probably be no problem. Of course, in a boat, you're likely to be the highest point in the water, and likely to attract a strike, so the best advice is to head into shore. The second-best advice is to dive into the water and stay under the surface.

Of course, fish can't get out of the lake when there is a storm, so they can't avoid the lightning. And fish don't react well to having an electric current pass through them. One of my colleagues, Katie Gilmour, who is a fish biologist, tells me that they take advantage of this when they want to sample fish populations. They use a technique known as electro-fishing to collect samples.

The way they do this is with an electrode at the bottom of the stream and another at the top. Through this, they pass quite large currents, one hundred amps or so. This zaps any fish in between and knocks them out for a few minutes. The same thing seems to happen at the surface of a lake during a thunderstorm. Biologists notice fish kills after lightning storms and, interestingly, they seem to be very species-specific. The one they see most frequently is something called a cisco, a surface-feeder that feeds off the plankton at the surface of the lake. So surface currents seem to be getting them.

MANOEUVRING MOLECULAR MINUTES

▲

How do you set an atomic clock, and what do you set it to?

Dr. Rob Douglas, Physicist with the Time and
Frequency Standards Group at the National
Research Council in Ottawa:

When you buy a brand-new atomic clock, or build one, it is not yet operating. So you put a battery in, and turn it on, the same as any old clock. Now all you have to do is set it. To do this, you turn to the best source you can find, which is another atomic clock. If you look back at this process over the decades, you find it peters out in 1958, when the first atomic clocks were set relative to the rotation of the Earth.

If all the atomic clocks were to fail at the same time, we would have to look into space for another reference tool. After atomic clocks, the most accurate timekeepers in the universe are pulsars. These are the spinning stars that are left over when a supernova explodes. They spin at a very constant rate and send out pulses of radiation. We know what the period of these rotations are for various pulsars, and could use them to measure the length of a second accurately. Knowing how long a second is would allow us to recalibrate our clocks. Despite the fact they're scattered across the galaxy, we can still monitor their rotation extremely well and they become the second standard after the atomic clock.

But while knowing how long a second is allows us to measure the passing of time, it doesn't tell us what the actual time is. We can't tell whether it is noon or one o'clock. So the trick that is used for different laboratories around the world to keep atomic time scales that are independent, and they also keep a coordinated time scale that is sort of a world average. Each of these clocks can measure local noon based on the Earth's rotation. When these clocks are compared, that generates the world average, which is reflected in the replacement for Greenwich Mean Time, now known as Coordinated Universal Time, or UTC, which forms the basis of time scales around the world.

STOP THE WORLD, I WANT TO GET OFF
▲

If the Earth stopped spinning would there be any significant gravi-tational changes?

Dr. Robert Mann, Director of the Guelph-Waterloo
Physics Institute and Professor of Physics and Applied
Mathematics at the University of Waterloo:

Right now, the Earth spins around its own axis at about 464 metres per second, or 1,670 kilometres per hour. If we were able somehow to stop the Earth from spinning, this would create worldwide havoc. If we stopped all of a sudden, the most noticeable effect would be that anything not firmly secured to the Earth's surface would immediately start flying off at the speed the Earth had been moving. The oceans, animals, people, and automobiles, for example, would all take off.

However, the velocity would not be fast enough to escape the Earth's gravity. To do that, you have to be going more than 28,000 kilometres per hour. So what would happen is all the people, animals, and things that weren't nailed down would rapidly fly off the Earth's surface, and then crash down some distance elsewhere. The oceans would slosh over the conti-nents. Buildings would not be able to stand the stress and would crack apart. The polar icecaps would undergo breaking, as though the ice were being slammed against a wall. It would be pretty horrific.

How far and fast things would fly depends on where they are on the Earth. Everything would fly in the same direction as the Earth's rotation. Since the sun rises in the east and sets in the west, we would all fly eastward, but we would fly at different speeds. At the Equator, we would fly at nearly 1,670 kilometres per hour, much faster than if we were close to one of the poles. This is faster than the speed of sound, so everything that flew off the Earth's surface near the Equator would do so with a sonic boom.

Beyond the initial catastrophes, there would be plenty of other effects on the Earth. The circulation of the winds would be radically altered. The tides, once everything had settled down, would also change, due to the loss of rotation, and the high/low-tide cycles would no longer exist. No living thing would survive, except, perhaps, bacteria.

Gravity itself would be unaffected by the stop. The Earth would continue to rotate around the sun. The day would now last a full year, since the Earth would no longer move its face in relation to the sun every twenty-four hours. So we would have six months of daylight and six months of darkness. Stopping the spin would not change the gravity felt by anything surviving on the planet, since the dominant effect is the attraction the Earth has towards its centre.

So stopping the Earth spinning would not have much effect on the gravity, but it would make for one heck of a messy day.

ROTATION REVELATION

▲

When watching spinning wheels in a movie, such as a spoked wheel on a motorcycle, why does it sometimes appear that the wheels are going backwards?

Dr. Fenella de Souza, Aerodynamics Lab at the
National Research Council in Ottawa:

You will not be surprised to hear that this is an optical illusion. It is a little complicated to explain, but imagine this: You're watching a movie and you see a motorcycle with spoked wheels going by. As you know, a movie is not a continuous picture. It is a series of pictures shown at twenty-four frames per second. Now, imagine the wheels of the motorcycle in the film are turning at exactly twenty-four rotations per second. Each time a frame of the movie was shot, the wheel would have turned exactly one rotation. It will appear as if the wheel is not turning at all. The wheel is turning at twenty-four revolutions per second and the camera is taking a picture twenty-four times per second.

Now imagine the wheel is going a bit faster than twenty-four rotations per second. With every frame the wheel will have rotated a bit more. It will look as if the wheel is rotating slowly forwards. Next, imagine the wheel is going less than twenty-four rotations per second. With every frame the wheel will have done slightly less than one rotation, and it will actually appear to be going backwards. This is similar to the effect of a strobe light in a disco. The strobe flashes at regular intervals and what you actually see is only what is going on when the light flashes.

Interestingly, you don't see this phenomenon only in the movies. If you ride a bicycle at night, for example, you might have noticed that your spoked wheels create the same illusion of either going forward very slowly, or even going backwards. The explanation for this is that the light from the street lamps is actually cyclic – it is flashing at about sixty hertz – sixty times a second – much faster than the eye can perceive. But if the frequency of the light flashing is not exactly the same as the frequency of your wheel turning, it is going to look like your wheel is going either slightly forward or slightly backwards.

This phenomenon is used in science and engineering to study anything that is cyclic. Say you want to look at something turning very fast, but you only want to look at this object when it is at a certain point in the cycle. For example, in aerodynamic research, let's say a helicopter rotor is turning at a certain frequency. Perhaps you just want to look at what happens when the rotor is in a certain position. If you have a strobo-scope flashing at exactly the frequency of rotation of your helicopter rotor, you can see what happens when the rotor is in that position. In a way, it is using light to slow down time.

HOT ENOUGH FOR YOU?

▲

We always hear about "absolute zero," in terms of temperature. But is there an "absolute hot" – a temperature at which matter, light, energy, and heat no longer exist?

Dr. Christine Wilson, Professor of Physics and
Astronomy at McMaster University:

The short answer to this question is: not as far as we know. In any conditions that we can create or theorize about, the limits to heat are essentially infinite. And while the conditions and form of the matter may continue to change, there is always still matter.

When you think about it, the most extreme heat you could ever possibly imagine would be the physical conditions like those in the early universe. We think the early universe started at the Big Bang, and at that time the whole universe was compressed into an infinitely small volume, and therefore was probably infinitely hot. It is hard to say exactly what the temperature was at different times in the early universe, but at about one-hundredth of a second after the universe was formed, the temperature in the photons and the particles would have been about a hundred billion degrees Kelvin.[*] That is a difficult temperature to fathom, but it is obviously pretty hot. But even at that temperature, it is still cool enough that protons and neutrons, the things that make up ordinary matter, could exist.

If you go much hotter than that, those atomic particles will break up into their individual constituent particles, the quarks. That happens somewhere above several trillion degrees Kelvin. Once the protons and neutrons have broken into quarks, then you have pretty much gone as far as you can go, in terms of

_____

[*] The Kelvin scale is a temperature scale whose zero point is considered the lowest possible temperature of anything in the universe. That point is known as "absolute zero," and is equivalent to –273.16 degrees Celsius.

our current understanding. This is really pushing theoretical physics a bit, but we think that once you are down to quarks, there is no limit to how hot you can get.

Perhaps, if quarks were found to have substructure within them – if we could break them down into more fundamental particles – the picture would change. Since we have not studied quarks sufficiently, we can't say this won't happen for sure, but our theories don't predict anything more fundamental than the quark. So as the temperature rises, they just get hotter and hotter and move faster and faster.

We are trying to study these conditions now in some of the very big particle colliders. By smashing together heavy elementary particles, like protons or neutrons, at huge speeds, we can momentarily make things hot enough to break them down into quarks and gluons, creating a "quark-gluon plasma."* But we are really just starting to get to that stage now, and we have a long way to go to push the universe's limits.

ALICE IN GRAVITYLAND

If you could drill a hole straight through the middle of the Earth, and if you slid down the hole you had drilled, would you just float weightlessly at the centre of the Earth? And if the slide were frictionless, would you go back and forth from one end of the tunnel to the other, coming out feetfirst at one and headfirst at the other?

---

* Gluon is a constituent of subatomic particles that has neither mass nor charge, and that holds together quarks.

Dr. Mark Shegelski, Associate Professor of Physics
at the University of Northern British Columbia:

The best way to look at this is to take it step by step. First of
all, let's simplify things and assume that there is no air resistance and no friction. We will also assume that you can drill
this kind of tunnel through the molten core of the Earth, and
somehow insulate it so that it isn't thousands of degrees
in the middle. Now you jump feetfirst into the tunnel.
Immediately, the force of gravity is going to pull you towards
the centre of the planet, and so you will start to accelerate and
move at faster and faster speeds towards the centre of the
Earth. By the time you reach the centre, about twenty-one
minutes later, you're going to be moving at an enormous
speed: 7.9 kilometres per second, or about 28,000 kilometres
per hour. So to answer the first part of the question, you are
not going to stop at the centre.

After you pass the centre, though, you will start to slow
down. More than half of the Earth is now above your head,
and its gravity is pulling back on you. So for the next twenty-
one minutes, you will slow down, and you will come to rest
just as your feet poke out of the hole on the other side of the
Earth. Of course, now you will start falling back, headfirst,
as the whole process starts again. Unless someone grabs you,
for the foreseeable future you will be yo-yoing back and forth
through this hole. In fact, this is a kind of motion that is quite
familiar to physicists. We see it in a lot of different everyday
experiences, like the pendulum swinging back and forth in a
grandfather clock. It is called simple harmonic motion.

If something were to slow you down, so that you stopped at the precise centre of the Earth, you would experience an interesting phenomenon. The force of the Earth's gravity would still be acting on you, but it would be acting on you equally from every direction. So the result is that the net force on you would be zero. So, yes, as the listener suspected, you would be weightless. You would hang in the centre of the Earth until someone figured out a way to come and get you.

FRICTION PREDICTION
▲

Why is it impossible to build a perpetual-motion machine? Why could an electric generator, for example, not run on its own power?

Dr. John Crawford, Professor of Physics at
McGill University:

The short answer to this question is the Law of Conservation of Energy, which basically says you can't win, and the Second Law of Thermodynamics, which says you shouldn't even make the bet.

The Law of Conservation of Energy demonstrates that you can't achieve perpetual motion because, in any system, you can't avoid losing energy. Most commonly the losses are to friction and heat. In the case of electrical motors and generators, we can make them really efficient, with high-quality windings and bearings, but there will still be a little bit of friction. So an electrical motor running on its own power will slowly run down, as energy leaks out into the surrounding environment in the form of heat.

Which is not to say you can't get pretty close to perfect efficiency and perpetual motion. It would, however, take an enormous budget. Here's my idea: To begin with, I'll need a spacecraft. It doesn't have to be big, but I'll need to get out to intergalactic space, where the density of matter is only about one hydrogen atom per cubic metre or so. Then all I need is a bicycle wheel, which I'll throw out the airlock with a good flick of the wrist to get it spinning. The bicycle wheel will continue to rotate nearly forever, so that would be perpetual motion. Unfortunately, even then there's a little bit of friction because that one atom per cubic metre would bounce off the wheel every now and then, and a little energy would be lost.

So that brings us to the Second Law of Thermodynamics, which is the one people usually forget when designing perpetual-motion machines. There's a reason for that, and that is because it sounds simple, but it is remarkably subtle, and so it can be difficult to understand. I will take a short stab at explaining it. The first part of the story is that, to do work, you have to take energy from a high-temperature reservoir and have it flow into a low-temperature reservoir. The second part of the story is that, in order to have 100 per cent efficiency, the low-temperature reservoir has to be as low as it can be, absolute zero – zero degrees Kelvin. The problem with that is that we can't get to zero degrees Kelvin. You can approach it, but you can never get there. So the Second Law of

Thermodynamics says nature won't let you have 100 per cent efficiency, which is what you need for perpetual motion.

JUMP WHEN I SAY JUMP

▲

Can you save yourself by jumping up just before a falling elevator hits the bottom?

Dr. Ron Lees, Professor Emeritus of Physics at the University of New Brunswick:

Jumping might allow you to lessen the impact a little, but if the elevator falls any real distance, you aren't likely to do yourself much good. The problem is, we can't really jump very high. Most of us, if we're really desperate, with our life passing before our eyes, might manage about half a metre. If you jump half a metre, then the speed at which you take off is going to be reduced from the elevator's speed when it hits. Unfortunately, to reduce your speed to zero, the elevator could only fall for a distance of half a metre.

Let's get a little more specific with an example. If you can jump up half a metre, then when you leave the ground, you would be going at just a little over three metres per second (about 11 km/h). If you fall from four metres, about the height of one storey, then you would be going at a little less than nine metres per second (32 km/h) when you hit the ground. So if you time your jump perfectly, you could reduce that to six metres per second (21.5 km/h). That would be the same as if you fell from a height of only one and two-thirds metres, which would not be too bad.

Now let's take that up to the height of three storeys, about twelve metres or so. If you don't jump, you'll hit the ground at fifteen metres per second (54 km/h), which would make for a pretty uncomfortable landing. If you do jump, you'll reduce your speed to twelve metres per second (about 43 km/h) on landing. That is still pretty fast – about the same as falling off a two-storey building – and the landing will still be pretty tough.

Of course the biggest problem with jumping in a falling elevator is the practical one of when to time your jump. You would need X-ray vision to know just when you're about to hit the ground, so as to time your jump properly. So, given that, and the impact you're still likely to feel if you fall from any significant height, don't hold out too much hope for saving yourself by jumping in a falling elevator.

HARMONIC HISTRIONIC

How can a sharp sound break glass?

Dr. Ian Cameron, Professor Emeritus of Physics at the University of New Brunswick:

The answer is based on the nature of sound. Sound essentially is a transmission of pressure pulses through the air. So, if the sound is loud enough and the pressure generated is high enough, and it strikes something reasonably fragile like a window, then it can certainly blow it apart.

Let me give you an example of how a sound pressure wave is generated. Suppose we take something like a vibrating tuning fork. Every time the prong of the tuning fork vibrates forward, it compresses the air in front of it. In other words, it pushes the molecules of the air together, creating a higher local pressure. Because the molecules of the air are essentially in continual contact with each other through collisions, this pressure pulse is transmitted through the air at a speed of about 330 metres per second. So, what we call an acoustic wave is a series of these pressure pulses.

When these pulses come along and hit glass, they literally push it. The air molecules in the high-pressure region physically strike it. If you have a very loud sound, like a big gun firing or a hand grenade exploding, the pressure rises very, very rapidly over the front of the wave, and so the pressure on the outside of the window rises almost instantaneously to a very high value. Since the pane is thin and fragile, this sudden pressure increase just blows it to bits.

Incidentally, I remember, as a very young kid during the Second World War, that sometimes when an explosion like that hit the windows, it threw them out rather than blowing them in. The reason is that this heavy pressure pulse of the explosion is immediately followed by a low pressure, which can actually suck the windows out rather than pushing them in.

When most people think about sound breaking glass, they imagine opera singers breaking wineglasses, and this is a slightly different phenomenon. Suppose you take a wineglass and you tap it. You'll hear it ring, and the tone or frequency of the sound will settle down to just a single note. This is known as the natural frequency of the object, and it will depend on the size and shape of the glass. An opera singer can break a glass if he or she can produce a sustained note at exactly the natural frequency of the glass. What happens is that the wineglass experiences something called resonance.

If you want to visualize this, you might consider yourself pushing a child on a swing. You know that if you want to generate a fairly large arc with a minimum amount of effort, you should always push at exactly the same rate as the swing itself is swinging. With every push the swing will go higher and higher, and the swing is resonating with more and more energy. With the wineglass, the pulses of sound pressure from the singer's voice act like a sequence of pushes on the glass, making it vibrate harder and harder. The glass just keeps taking energy from the singer's voice, vibrating with more and more energy until the structure of the glass just can't take it any more, and the glass blows itself to bits. It is hard for anyone but opera singers to do this because, first of all, they can sing very loudly and put a lot of energy into the glass, and second, they can maintain exactly the right pitch to stay at the natural frequency of the glass. Of course, they are pretty good at holding a note for a good long time as well.

A Matter of Perspective

▲

Why do objects appear to get smaller the farther away they are?

Dr. John Samson, Professor of Physics at the
University of Alberta:

This is a question that, at first glance, looks simple, but the more you think about it, the more complicated it gets. To start, and this might come as a surprise to some, it is not always true that objects look smaller when they are farther away. Most of the time it is true, and the reason is that light generally travels in a straight line. The exception to this rule is where things get interesting.

So let me explain the more common case, in which objects look smaller as they get farther away. The best way to do that is to think about looking at a light bulb. Light from that bulb goes in all directions in straight lines. We see that light with a rather simple optical device called an eye. The eye is just a lens with a screen behind it, which we call the retina. Light travels to the lens in a straight line, and the lens bends it so that it focuses it to a point on the retina – and so we see one light bulb. Next, put two light sources out there, for example, the tail lights of a car. Light comes out of each bulb, going in straight lines. The light from each bulb hits the lens at slightly different angles and is focused to two points on the retina. Now, imagine the car drives farther away. If you draw a diagram, you will see that when you draw straight lines from the tail lights to the lens of the eye, the angle of the light sources to the lens of the eye will be reduced the farther away the car is. The light will still be bent by the lens, but the points on the retina will get closer together. Your brain will perceive that closeness as the shrinking size of the distance between the two tail lights – the car will look smaller. Eventually, the lines from the lights will be effectively parallel and the eye will take those two points of light and focus them on one point in the eye. Nothing can be smaller than a point, of course. So, that is how things appear smaller as they move into the distance.

Let's now look at the exception to this rule, with a bit of an experiment. Fill a glass with water. Now take a fork in your hand and look at it, nice and close. Next, look at the fork through the glass of water at a greater distance. The fork will look bigger, even though it is farther away. You probably think I am cheating here, because we have basically used the glass of water as a magnifying lens, and that is quite true. This is not just a trick, though. In space, light rays get bent all the time and often do not go in a straight line. Einstein's Theory of General Relativity explains how large masses, such as black holes, can bend space and light, achieving the same effect as looking at a fork through a glass. Astronomers have actually used this "gravitational lensing" to get closeup looks at distant stars, by looking for stars or galaxies that are in line with each other, when viewed from the Earth. The light from the farther star will be "magnified" by passing by a nearer galaxy or star on its way to us. In that case, some rather interesting distorted images of the far star are generated.

PLEASE NOTE THIS NUMBER

▲

Why has the musical note A been set at 440 hertz? Was it arbitrarily chosen or is there some reasoning behind it?

Dr. Lynn Cavanagh, Assistant Professor in the
Department of Music at the University of Regina:

There are some good reasons why A has been set with 440 hertz as the standard, and there's quite a lot of history associated with it.

The first thing to note (so to speak) is that it wasn't always the case that there was a standard. It is actually an innovation of the nineteenth and twentieth centuries. In Europe in earlier centuries, the pitch of any note varied widely from place to place and century to century, both up and down. In Germany alone, prior to 1600 the pitch of the A above middle C ranged from 567 hertz, in one place and time, to as low as 374 hertz in another.

There were two reasons for this. Firstly, standardization wasn't necessary because composers and performers took the variation of pitch into account. Secondly, local pitch was usually based on the organ in the local church. These organs weren't easily retunable, and weren't built to any standard, so the construction of the organ would determine the pitch that singers would sing, and how the other instruments would be tuned. Travelling organists who were playing from a score that had been composed with a different pitch in mind simply transposed the score by sight, as they were playing, to accommodate the other instrumentalists and singers. These organs were also very temperature-sensitive and would change their fundamental pitch over the seasons, or even from day to day. Given that, maintaining a standard pitch would have been terribly difficult, and there would not have been any point to it.

In the nineteenth century, however, a trend began in Europe that started inexorably raising the pitch level of performances. Audiences were becoming larger, they were listening in large concert halls and opera houses that could accommodate brilliant, higher pitches, and the organ was no longer quite as standard an instrument. Wind-instrument makers, eager to find a niche in the market, responded by developing and selling instruments with an ever-slightly-higher fundamental pitch. The musicians no longer transposed the sheet music down as organists had done, so the pitch of performances just kept getting higher and higher.

It got to the point where, in some opera houses, the standard pitch for A was at 450 hertz, and singers were struggling to sing a work that had been written with A at 423 hertz in mind – and that was definitely a problem. It was damaging singers' voices, and they went to their managers and insisted that something be done.

In 1859, the French government, always active in matters of culture, took the lead and commissioned a study. It recommended A should be set, by law, at 435 hertz. This was a compromise solution, arrived at because singers needed the current standard to come down, but audiences wanted it as high as possible to maintain the brilliance of sound.

A at 440 hertz was established as an international standard in the 1930s, primarily because of the needs of the broadcasting industry. Broadcasting was becoming international, and it was inconvenient for orchestras and pianos in different places to be tuned to different pitches. Over several years, "A-440" was gradually agreed upon, and it was confirmed as the official international standard at a conference in London in 1939.

REFLECTING UPON LIGHT

▲

If light were introduced into a perfect, hollow spherical ball that was lined with a perfect reflective surface, would there be light inside that ball for as long as the ball existed? And if light continued to be introduced, would the ball eventually explode?

Dr. Mark de Jong, Project Leader of the Canadian Light
Source Project at the University of Saskatchewan:

This is a good question, and the simple answer to the first part of it is, yes. If there were no absorption by any of the walls, and the light kept on reflecting around and around, it could continue bouncing forever. If there were a way to introduce more and more light as you went along, then you would build up energy in the ball, in the form of light. An interesting result of this would be that, because of the equivalence of mass and energy, which Einstein so famously pointed out, the ball would actually start getting heavier.

The answer to the second part of the question, though, is no, and that is interesting, too. Remember that one of the conditions here is that the inside of the ball would have to reflect perfectly. The only way that an explosion could occur is if something inside the ball, or the ball itself, started absorbing energy and heating up the material until it came apart in an explosion. But because there is perfect reflection in this ball, there can't be any absorption, so there can't be an explosion. The light just stays in the ball in the form of light.

Of course, this kind of device would be very difficult to set up. The biggest problem is that little hole you would have to make, in order to keep pumping more light into the ball. It would, of course, also allow light to leak out.

We have used similar sorts of principles in a lot of our modern technology. One example is the laser. In many standard lasers, you create a little cavity by putting two mirrors facing each other, and having the light bounce back and forth between them. Then you actually boost or amplify this energy with a laser medium placed between the two mirrors. It isn't perfectly efficient, though, and the amount of light you get out of the laser is just a very, very, small fraction of the total amount of energy bouncing around inside it.

Another example is the optical fibres we use in communications technology. They have an extremely reflective coating, so that light doesn't escape through the walls of the fibre. The light is confined in the fibre, bouncing from wall to wall, until it comes out the other end of the fibre at its destination.

LIGHT AS A FEATHER

▲

If a flock of birds is loaded into a plane and during the flight, they take off and start flying around the cargo hold, does the weight of the plane change?

Dr. Mariana Frank, Professor of Physics at
Concordia University:

The answer to this question depends on whether the birds are accelerating vertically while they are flying. That is because Newton's Third Law governs the effect of the birds on the plane. This law is known as the Law of Action and Reaction.

If we ignore acceleration, then the situation is fairly simple. While the birds are stationary, their weight pushes down on the floor of the airplane. The plane amounts to its own weight plus the weight of the birds. Now we need to consider what happens if the birds are flying. If the birds are hovering in the air, their weight acts as a downward force on the air below them. So the air pushes down against the floor of the plane with an additional force equal to the bird's weight. So as far as the weight of the plane is concerned, the total force hasn't changed. The effective weight is the same.

That is what happens if the birds are just hovering, or flying forwards and backwards in the plane. If they are accelerating upwards or downwards, then the situation changes. If the birds are accelerating upwards, they are pressing the air below with the force of their weight, plus some additional force to help them move up, and all this is directed towards the bottom of the plane. So the effective weight of the plane has been increased by this new force, the mass of the birds multiplied by the acceleration of their movement. For a moment the plane is going to be heavier.

Of course, if the birds accelerate downwards the opposite effect will happen, and the plane will become lighter. But don't forget, this effect won't happen for very long, because the birds can't move up or down very far inside the cargo hold of a plane. Realistically, unless all the birds suddenly moved to the ceiling at the same time, or plunged simultaneously to the floor, it would be very hard to see any effect.

7

Sizzling Suns and Magnificent Moons:

SPACE SCIENCE AND ASTRONOMY

BABY, IT'S DARK OUTSIDE

▲

Considering that there are tens of thousands of light sources in space,
why does the night sky appear dark?

Dr. Harvey Richer, Head of the Gemini Telescope
Project in the Astronomy Department at the
University of British Columbia:

This is a wonderful question and it has a very long history. To
understand the answer, let's use a simple analogy. If we look
into a forest, where there are lots of trees, eventually our line
of sight is going to hit one of the trees and we won't be able to
see through to the other side of the forest. This is the kind of
puzzle we are dealing with when we look at the night sky. It
is a puzzle that is called Olber's Paradox.

Heinrich Olber was a German astronomer. He assumed
that we have an infinite universe and that it is populated

universally with stars. He said that our line of sight ought to hit a star wherever we look, and hence the Universe should be as bright as the surface of a star. But Olber's assumptions turned out to be incorrect.

The universe isn't infinite. While different astronomers dispute its exact age, they do generally agree the universe has an edge. So, if there is not an infinite number of stars, that means there are going to be gaps where there aren't any. So there is no reason why our line of sight necessarily has to hit an individual star.

The situation is even helped by the fact that galaxies have evolved. The galaxies, and the stars in them, are even younger than the universe itself, so in the earliest phases of the universe there weren't any galaxies producing stars and light. When you look very far back, near the edge of the universe, where the galaxies haven't formed, there is no light to intercept our eye.

There is a second, minor effect, caused by the expansion of the universe. Because the universe is expanding, the light from the more distant galaxies is red-shifted. Red shift is an effect we see when we look at stars and galaxies moving away from us. The light and energy from the stars gets stretched out and arrives here a lot weaker than when it set out. So we see a lot less energy than we would expect if the universe weren't expanding. This helps to make the night sky black, but it is not the main contributor.

HOLDING IT ALL TOGETHER
▲

We know the forces which hold the Earth in its orbit are related to the masses of the Earth and the sun. We also know the sun is losing mass as it radiates energy. Is that loss of mass affecting the Earth's orbit around the sun?

Dr. Peter Bergbusch, Professor of Physics and
Astronomy at the University of Regina:

The sun is losing a fraction of its mass, but it is a very small amount. It is losing somewhere between one tenth of a trillionth and one hundredth of a trillionth of its mass every year. It is a lot of stuff, but in terms of the entire sun, it is not a great deal of matter. On the other hand, the sun is also gaining mass. Dust and comet remnants and that sort of thing fall into the sun. But the rate of mass outflow is greater than the mass inflow.

This loss of mass does have a gradual effect on the Earth, but it is very small. As the mass is lost, the force of gravity holding Earth and the sun together is slightly reduced, and the two bodies move apart. Since the formation of the Earth about five billion years ago, its orbit has increased in radius between thirty and forty thousand kilometres, which is about three times the Earth's diameter. And since the orbit is longer, that means the year is longer, too. But this hasn't made a lot of difference. Assuming the calculations are all correct, our current year is only about eight hours and twenty-five minutes longer than it was when the planet first formed.

The loss of the sun's mass hasn't had a huge effect on us. You'd have to have a huge change in the mass of the sun to radically affect the position of the Earth in its orbit. You'd almost have to make the sun go away before we would go flying off into space.

Turn Left at the North Galactic Pole

▲

I think of our galaxy as a giant disk, flying through space and spinning at the same time. I know our solar system is fairly close to the outer edge of that disc, but exactly where are we? If the galaxy is travelling in a northerly direction, are we on the eastern edge or the back, or is the bulk of the galaxy following along behind us?

Dr. Peter Bergbusch, Professor of Physics and
Astronomy at the University of Regina:

This is a tricky question to answer, because the first thing you have to decide is how you are going to measure your motion. For example, if you are driving down the highway in your car and you decide to measure your motion against other cars on the highway, you can get very different answers depending on whether they're coming towards you or moving in the same direction as you. That is why we normally measure our speed with respect to the road itself. That is what we call the "rest frame" of the car, since the road is at rest, or standing still, and you can measure your speed against that frame of reference.

But there aren't any roads standing still in space. As we look out, we see galaxies moving away from us in all directions due to the general expansion of the universe. But the galaxies also interact gravitationally, so the motion we observe is due to the combined effects of the cosmological expansion and of the effects of the force of gravity acting between them. We have to try to find something in the universe that we can believe is in the rest frame, and the closest we can come to that is what we call the microwave background radiation. It originates with the Big Bang, and as the universe expands, that radiation participates in the general expansion of the universe.

According to measurements made in relation to the background microwave radiation, our galaxy is moving through the universe at about 600 kilometres per second. As well as moving with the expansion of the universe, our galaxy is also spinning. Viewed from the North Galactic Pole, it would appear to be a disc rotating in a clockwise direction. The solar system is located a little less than two-thirds of the way out from the centre of the disc. If the front of the disk (taken to be in the direction of our galaxy's motion relative to the cosmological background radiation) is thought of as twelve o'clock, we would be at about four o'clock on the dial. That is, we are on the side of the disk that is rotating just slightly away from the direction of the galaxy's motion.

Our sun takes about two hundred and fifty million years to complete one rotation around the centre of our galaxy. So if you wait around for about one hundred and twenty-five million years, we will have moved to a point where we are on the front side of the galaxy.

Budda-Big, Budda-Bang

▲

If the universe originated in a Big Bang, as astronomers claim, then where is the point in space where everything began? Where would we look for it today?

Dr. Jaymie Matthews, Professor of Astronomy at
the University of British Columbia:

It is hard to understand, but there isn't one point in space where our universe began. That is a common misconception people have about the Big Bang Theory. There is no X that marks the spot where the Big Bang occurred. The Big Bang would not have been an explosion of matter into space, like a bomb going off, with the material flying outward like shrapnel from some central location. Rather, you must think of it as an explosion of space itself, with all the matter being carried along with it.

This is difficult for humans to visualize because, by nature, we are four-dimensional creatures. We perceive only three dimensions of space and one dimension of time, and the expansion of the three spatial dimensions is happening with us as part of it. It is impossible for us to have an outside perspective, because there isn't an "outside" from which we can look at the universe. The universe isn't necessarily expanding into anything. Everything we can possibly perceive is in those four dimensions. In trying to depict the Big Bang, astronomers and the media often resort to images of an explosion you can watch from a distance; but, in fact, you could never have that perspective from within our known 4-D universe.

The idea that there must be a point from which everything expands might seem to be reinforced by the early observations of Edwin Hubble. He discovered that no matter which way we look, other distant galaxies are flying away from us. So it is natural to assume that we must be at the centre of the expansion. But Hubble also found that the farther out you go, the faster galaxies are moving away from us. This is not what would happen in a normal explosion of matter in space. The only way to explain it is if the geometry of space itself is actually expanding and carrying matter along with it. In that case, no matter where you are in the universe, you would see exactly the same perspective. If we could go to a galaxy five billion light-years away from us and look back towards the Milky Way galaxy, it would look as if the Milky Way and all other galaxies were flying away from us in exactly the same way it looks from here. Everyone looks like they are in the centre. It is a very democratic, egalitarian universe we live in.

Even so, it is still hard to picture an expansion without a centre. It might help to imagine going back in time, to an earlier point in cosmic history. Back then, the geometry of the universe would have had a smaller scale compared to today, but if you looked around you, it would still look immensely vast. There would be no physical edge to it. You wouldn't feel like you were squeezed into some sort of box. Ten million years after the Big Bang, the scale of the universe would be at least a thousand times smaller than today, but you would be able to see light arriving from parts of the Universe about ten million light-years away.

However, you would notice the universe, on average, was denser and hotter than it is today. If you went back to a time when the universe was only a billionth of a second old, then light would only have had the chance to travel a tiny distance (like in a thick fog). So hypothetical time travellers (if they could survive the extreme conditions) would not even be able to see their own toes. Their bodies would stretch beyond the observable "horizon" of the young universe. Today, our horizon extends out for billions of light-years, but still does not define a physical edge to the universe.

THE REPERCUSSION OF A
LUNAR CONCUSSION

▲

Is it possible for a comet or asteroid to hit the moon? What would be the consequences on Earth?

David Balam, Research Assistant in the Department of
Physics and Astronomy at the University of Victoria, and
Principal Observer of Space Guard Canada, a program
that tracks near-Earth objects:

Large impacts on the moon do happen. There was one recorded in 1178 by Gervaise of Canterbury, a chronicler of the time. He interviewed five Canterbury monks who described this scene: "There was a bright moon and, as usual in that phase, its horns were tilted towards the east. Suddenly the upper horn split into two. From the midpoint of the division, a flaming torch sprung up, spewing out fire, hot coals and sparks." We believe this is the first actual report of a lunar impact. The object that hit the moon must have been between fifty and one hundred metres in diameter for the monks to see it here on Earth with just their naked eyes.

So, impacts can and do happen on the moon, and the result would be fairly significant. Beneath the impact site, there would be a great compressed plug of material, and this would rebound explosively. Any material that attained a velocity greater than the escape velocity of the moon would be thrown out into space and not fall back. Any material that was travelling below escape velocity would go into orbit and eventually impact the moon again. But there wouldn't be much effect on Earth. If anything, there might be some dust, but that would filter very slowly into the upper atmosphere.

The meteorite seen by the monks is much too small to have had an indirect effect on Earth. For that, the moon would have to be hit by an object measured in kilometres across. If something that large hit the moon, our main worry would be dust. A large impact would send out lots of fine dust that could rain down on the Earth. This would tend to block out sunlight, and, if it lasted for several years, then it could affect farming, leading to economic downturns. There are objects that size out there, and we are tracking several thousand asteroids right now. One we tracked in 1998 was a bit more than seven kilometres in diameter. That is pretty big, considering the object that could have killed off the dinosaurs was in the ten-kilometre range. Some of these objects have come pretty close to hitting the moon, but they only hit the moon about once every ten million years. Even smaller objects, in the ten-metre range, only hit the moon once every hundred thousand years. So, we haven't got much to worry about.

Planetary Spin Cycle

▲

All the planets, except Uranus, orbit the sun with their poles pointing up and down. But Uranus spins on its side. Is this because of some kind of cataclysmic collision?

Dr. Norman Murray, Astrophysicist at the Canadian Institute for Theoretical Astrophysics at the University of Toronto:

The short answer is we don't know why it spins on its side. There are several suggestions, but only two seem reasonable to me.

The oldest suggestion is that Uranus suffered a giant collision, which tilted the spin axis of the planet. It would have had to be an enormous impact, and one might expect it would have done other things to the planet. For example, it would have changed the shape of the planet's orbit, and that doesn't seem to be the case. The orbit of Uranus is more or less circular and the giant impact should have changed it to more of an ellipse. So that has led most people to believe such a giant impact did not occur. Recently, however, it has occurred to me and others that the interaction between Uranus and a population of smaller bodies, such as those that were ejected from between the planets to form the Oort cloud, would act to decrease the eccentricity of Uranus's orbit. That makes the idea of a giant impact more plausible.

The second suggestion is not that the planet's spin axis has changed, but that its orbital plane has. The planets are believed to have all formed out of a giant gas disc, and it's possible the plane of the whole disc has tilted. If Uranus had already formed before the other planets, and was orbiting around inside this gas disc, its spin axis would be pointing in one direction and would have stayed that way as the disc began to tilt. Uranus's axis would have remained constant, but its orbit would have moved along with the disc. The other planets, forming after the disc had tipped, would have a new orientation, lining up with each other but at a different angle from Uranus. But the problem with this theory is that there is no evidence Uranus formed before the other planets. The big question is that, if Uranus has an odd spin, then why don't all the other planets have funny spin axes? In fact, the rest of the giant planets, Jupiter, Saturn, and Neptune, all have spin axes that more or less point north and south.

TACKING IN A LIGHT BREEZE
▲

How would you tack or point a solar sail in space?

Dr. Kieran Carroll, Manager of Space Projects
at Dynacon Enterprises in Toronto:

A solar sail is basically a very large mirror. Any mirror will act as a solar sail by reflecting the photons that come off the sun, but to be practical as a sail, the mirror has to be about the size of a football field, and very thin. The types of mirrors people have been looking at use the same kind of technology that is used to make potato chip bags. Look inside a potato chip bag: you'll see it is plastic backed with aluminum. If you were to cut one of these open and lay it flat, you would see it makes a pretty good mirror. Then, if you were to take the plastic from the bag and make it much thinner – as much as twenty times thinner – you could roll that up into a very compact roll that would fit on a spacecraft. Then, once you were in space, you could deploy the bag to form your huge solar-sail mirror.

A solar sail works by taking advantage of the nature of light. Light is composed of packets of energy, called photons. Each of these photons has momentum, just like a baseball has momentum when you're throwing it between the pitcher and the catcher. When the catcher catches the baseball, he feels a force against the glove. Well, whenever a photon bounces off the mirror, the mirror feels a force from that photon. It is a very tiny force, but there are lots of photons coming off the sun, and their forces all add up to generate a large enough force to change your orbit.

Changing the orbit of a spacecraft using a solar sail relies on gravity as a kind of solar keel. Imagine your ship is in a circular orbit around the sun. One side of the ship will always be pointing towards the sun. If you speed the ship up, it will move outward and establish a new orbit a little farther away from the sun. Slow the ship down and it will move into a lower orbit. So, instead of using rocket thrust, a solar sail does the same thing with photons. Remember that one side of the ship is pointing towards the sun. If you put up your solar sail with the shiny side towards the sun, the photons will start bouncing directly off. Now, tilt the mirror so the photons bounce off backwards and that will push you forward. Since you're speeding up, that'll make your orbit bigger. Then, if you rotate the mirror around so you're bouncing the photons off forward, you slow down, and go into a smaller orbit.

Solar-sail technology is close to being a practical reality. You might see a solar-sail mission coming out of NASA within the next five years. There is a program proposal to put up a solar storm-monitoring mission, which would use a solar-sail spacecraft to hover closer to the sun than you'd get otherwise. In addition, the Canadian Space Agency, with my company, has been working to get some solar-sail technology development going here in Canada. So you may see a Canadian solar-sail mission some time in the next five to ten years as well. A sort of giant maple leaf in space.

LOST AT "C"

▲

If one were to travel vast interstellar distances at a rate faster than the speed of light, would we be able to see much along the way?

Dr. Ann Gower, Professor of Physics and Astronomy
at the University of Victoria:

According to the laws of physics, as we understand them, it is not possible for us to travel at the speed of light, let alone faster, so there isn't really an answer to this question. What can be answered is what you would see if you travelled through the galaxy at very close to the speed of light.

First of all, looking out the front window, any light from stars in front of you would be shifted to shorter wavelengths, or blue-shifted, because you're moving so fast. If you were going very fast indeed, really close to the speed of light, the light would be shifted out of the visible range completely, into the X-ray or gamma-ray wavelengths. Very high energy radiation would be hitting you very hard from straight ahead. It would be really dangerous, and you would likely be baked!

But whereas the light from stars in front of the ship would be blue-shifted, the opposite would happen behind you. The light from those stars would be red-shifted to longer wavelengths. These would be below our visible range, right down into radio wavelengths, so they would be invisible to the naked eye. This means the sky behind us would be dark, without much to see.

When you looked out the sides of the ship, you would also see very little. There is a very dramatic effect on the geometry of space, due to relativity, when you travel close to light speed. If you're going very fast, the stars will appear to be bunched close together in front of you. Everything will be compressed into a cone ahead of you, in the direction you're moving. It is rather like driving in a shower of rain. When you drive through a rainstorm, it looks as though all the drops are coming from the front. The same effect would happen in a spaceship as you approached the speed of light, but it would be much more extreme. All the stars would appear to be in front of you. So you're not going to see much out of the side windows.

Light waves aren't the only thing affected by travelling close to the speed of light. The universe is filled with cosmic background radiation, left over from the Big Bang. Like the light, the energy from the cosmic radiation would appear blue-shifted in front of the ship, making it much hotter and more energetic. It would become another source of X-rays raining down on the ship. Added to the light from the stars, it would create a very high energy situation!

Another form of energy, which we know must exist throughout the universe, is gravity waves. We haven't yet detected them, but we know they are there. These will also be intensified. As you travel closer to the speed of light, you would feel them as stronger and stronger bumps. So, even if you survive being baked by the shifted starlight and background radiation, you are liable also to be shaken violently by intensified gravity waves. Perhaps we could call it "shake and bake" travel!

NORTH TO POLARIS

▲

I was recently watching the movie Contact, *starring Jodie Foster, and my attention was drawn to the compass she had brought on her intergalactic voyage. I was wondering which way a compass needle would point in space, outside of the Earth's magnetic field? What would happen if it were near another planet?*

Dr. Douglas Beder, Associate Professor Emeritus in the
Physics and Astronomy Department at the University
of British Columbia:

I think Jodie Foster must have had great presence of mind
to find the time to look at her compass. I would scarcely be
able to catch my breath under those circumstances. But the
best answer to the question is that it depends. If you were
near a planet or near a star, you'd possibly be in a magnetic
field that was comparable to what we have here on Earth. In
that case, the compass needle would point in the direction the
field was pointing. But which direction that indicated would
depend very much on where you happen to be with respect
to the rotation of the star, and which way you're holding
the compass.

You would also have to be fairly close to the planet or star.
Magnetic fields drop off very rapidly as you move away from
the source. If you're more than a few times the diameter of the
planet from the surface, the magnetic field will have dropped
to about 1 per cent of what it is at the surface. So it might not
be strong enough to move the needle of the compass (unless
you had a perfectly frictionless compass and lots of patience
to wait for its excruciatingly slow needle motion). For example,
just thirty thousand kilometres away from Earth, whose radius
is about six thousand kilometres, the magnetic field would
have decreased so as to make a normal compass useless. Of
course, stars are much, much larger than the Earth, so you
would still find a reasonable magnetic field several million
kilometres away from a star like our sun, but you would be
rather hot at such a location!

On the other hand, neutron stars, also called pulsars when we detect their radio pulses, are exceptional in many ways. They have near-surface magnetic fields about a trillion times stronger than the Earth's magnetic field, and you would still feel an Earth-size field at a hundred thousand kilometres' distance. The problem with that location would be the exceptionally strong gravity field, about a thousand times Earth's surface gravity.

Farther away from planets or stars, you would be in the general magnetic field of the galaxy, but it is not a uniform, coherent field that is always pointing in the same direction everywhere. It is more influenced by local accumulations of dust in regions where stars are forming. Basically, it is just not good for finding the north or south pole of the galaxy.

Between galaxies, what would happen is a bit mysterious because we don't have good information about the amount and kind of matter out there. But a good estimate is that any magnetic field out there is likely to be even less than one millionth of the strength of the Earth's magnetic field, and again your compass wouldn't be much use.

CORONA CONUNDRUM
▲

The sun's core burns at a temperature of millions of degrees Kelvin and the solar atmosphere, or corona, is similarly millions of degrees hot. But between these searing regions, the sun's surface maintains a balmy temperature of only several thousand degrees. What physical process or mechanism keeps the surface cool? Is it a Lennox or am I a lummox?

Dr. Christine Wilson, Professor of Physics and
Astronomy at McMaster University:

In some sense, the question is a bit backwards. The question really ought to be: why is the corona so hot? After all, when you think about it, it is fairly easy to understand why the surface of the sun is cooler than the centre of the sun. The structure of a star like our sun is a balance between the competing forces of gravity, which wants to make it collapse into a very small object, and pressure, which tends to want to puff it out. The competition between those two forces creates a situation in which the inner parts of the sun are very hot: that is where you have fusion going on and the sun's energy is generated. As you move out from this hot centre, the temperature drops off. Eventually, you get to so far out that you don't have fusion any more. The energy just propagates out through the sun, and it gets colder and colder. So that is why the surface is cooler than the centre in a star.

The corona doesn't follow this nice logical picture. The corona is the outer atmosphere of the sun, extending outwards for millions of kilometres, and it can indeed reach temperatures of millions of degrees. To be perfectly honest, astronomers who study the sun don't really know how the corona can be so hot, but we have some clues. First of all, the corona is very thin. It is about ten billion times less dense than the atmosphere of the Earth at sea level. Because it is so thin, small amounts of energy can heat it very hot. The problem is to figure out where that energy is coming from.

The best theory now is that the energy that heats the corona comes from activity on the surface of the sun. Magnetic fields in sunspots can produce flares, shooting material off the surface of the sun. You may have seen photos of these stormy loops and whorls of material being blown out into space. We think that some of that energy gets transposed into the corona, perhaps by electric fields caused by the magnetic fields twisting up. The details of that are something that solar astronomers are still trying to work out, but we are pretty sure there is a connection because how the corona looks depends on the degree of surface activity on the sun. When there is a strong period of activity on the sun, the corona is much bigger than when there is a minimum of solar activity.

LOSING LUNA
▲

I'm familiar with satellites and space debris deteriorating in orbit and falling back to Earth. But why isn't the same thing happening to the moon? Why isn't it getting closer to the Earth? Shouldn't the gravitational and frictional forces between the Earth and the moon cause the moon to slow down?

Dr. Peter Bergbusch, Professor of Physics and
Astronomy at the University of Regina:

Satellites and space debris in low Earth orbit do fall to Earth, but mostly because of friction with the Earth's atmosphere. Although the atmosphere is very thin out at a couple of hundred kilometres, it is still dense enough to cause things to slow down and fall out of orbit. Objects well beyond the Earth's atmosphere, however, such as more distant satellites and the moon, don't experience this friction and so they don't slow down and fall to Earth.

As for gravity, it isn't causing the moon to slow down and drop towards the Earth, it is actually working to accelerate the moon in its orbit and push it away from the Earth.

This is a bit difficult to understand, so let me try to build a picture of the complex interaction between the Earth and the moon. Imagine yourself looking down on the Earth-moon system from above the Earth's North Pole. The Earth is rotating in a counter-clockwise direction. The moon is orbiting the Earth, also in a counter-clockwise direction, and the gravitational attraction between the Earth and the moon is causing the water on the Earth's surface to bulge a little towards the moon. This is the phenomenon we all know quite well as the tides.

However, because the Earth itself is rotating faster than the moon is orbiting around the Earth, an interesting thing happens. That tidal bulge of water, which the moon raises, is actually pushed forward a little ahead of the line joining the Earth and the moon. In effect, the tides are raised by the gravity of the moon, and then whipped ahead a little by the speed of the rotation of the Earth.

We now have to see how that, in turn, has an effect on the moon. The important thing to understand is that the tidal bulge is actually attracting the moon as well. As the bulge is rotated around by the Earth, it drags the moon along with it a tiny bit. And that bit of extra force acting on the moon increases its orbital angular momentum – its speed in orbit. And when you increase the speed of an object in orbit, it responds by increasing its distance from the object it is orbiting.

So, the same interactions that raise the tides on Earth are gradually pushing the moon farther away from the Earth. The amount of energy being passed back and forth here isn't huge, and we need not worry too much about losing the moon. It only moves away a tiny distance every year, and it would take hundreds of billions of years for it to move so far that it escaped Earth's gravity. Of course, our solar system won't be around for that long, since our sun will run out of fuel and turn into a red giant in about five billion years or so.

TENNIS ON THE MOON, ANYONE?

▲

Does a ball dropped on the moon bounce higher or lower than the same ball dropped from the same height on Earth? Is a parachutist better off to have his parachute fail on the moon or on Earth? What would the relative heights and speeds be for terminal velocity?

Leah Braithwaite, Scientist with the Space Life Sciences Program at the Canadian Space Agency in Ottawa:

There are a couple of questions here, and a few of us at the Space Agency put our heads together to come up with answers, and we will take them one at a time.

The answer to the ball bouncing question is, surprisingly perhaps, that a ball dropped on the moon should bounce to pretty much the same height as it would on Earth. That is thanks to the Law of Conservation of Energy. When you drop a ball on Earth, it is being attracted to the Earth by a force of gravity six times that on the moon. So by the time it hits the ground, it will be travelling a lot faster than it would be if it had been dropped on the moon. Therefore, more energy will go into the rebound of the ball from the ground on the Earth than would happen on the moon. The problem for the Earth ball happens on the way back up. It has a great deal of energy as it bounces back, but it has to fight that strong Earth gravity. A ball bouncing on the moon might have less energy in its bounce, but it will be bouncing up in the moon's one-sixth gravity, and so will bounce to the same height as the ball on the Earth.

This is a simplified scenario, since we are not worrying about things like air drag on Earth, and whether the ball responds differently to different compression forces, but we think those differences would be negligible for a ball dropped from shoulder height.

There would, unfortunately, be a big difference between the Earth and the moon for a parachutist. The biggest difference between parachuting on the moon and parachuting on the Earth is that every time our skydiver parachutes on the moon, his parachute is going to fail. That is because parachutes depend on air resistance, and there isn't any on the moon. On the one hand, the end result of skydiving with a failed parachute is going to be the same on the Earth or the moon. You're going to go *splat*. On the other hand, they might name the crater you make on the moon after you.

This leads into the last question, about the different terminal velocities on the Earth and the moon. Once again, terminal velocity is something that you encounter when you're falling and the air resistance that you're pushing through equals the force that is pulling you down. When you first jump out of a plane, you will travel faster and faster until, eventually, the air resistance will slow you down to the point where you travel at a constant velocity. For a person falling on Earth, terminal velocity is about two hundred kilometres an hour, and it takes about ten seconds and about 350 metres of falling to get to that speed. Unfortunately, since there is no air resistance on the moon, there is no terminal velocity. After you jump out of your spacecraft, you would continue to accelerate towards the surface of the moon, and you would just keep going faster and faster, until you stop rather suddenly.

WHY SO SAD, MR. MAN-IN-THE-MOON?

▲

Every two or three years there will be two full moons in a month, and the second one is usually called a blue moon, even though it is not really blue. But can the moon ever really be blue?

Dr. Peter Bergbusch, Professor of Physics and
Astronomy at the University of Regina:

There's a short answer here, and that is that, if the circumstances are just right, you can indeed have a really blue moon.
After certain volcanic eruptions, or a very fierce forest fire, the
atmospheric conditions can exist to produce a blue moon.
The secret is that you need dust or smoke particles of just the
right size and refractive index to be put high into the atmosphere. What happens is that light from the moon hits these
particles, and the particles act like mirrors that reflect the red
light off in all directions. The blue light, which is at a different
wavelength, passes right through the dust unaffected. Since
all the red light is gone, and only the blue light gets to us on
the ground, we end up seeing a blue moon.

This can even happen with the sun, so that we get a blue
sun. However, this is caused by exactly the opposite of the
effect that gives us a blue sky. The atoms and small particles
of dust in the atmosphere scatter blue light coming in from
the sun, and so we see blue light coming from every part of the
sky. The difference in the case of a blue moon or sun is the size
of the particles in the atmosphere, which scatter red light
instead of blue.

A "real" blue moon definitely is a rare occasion. The one
that I know of for sure occurred in 1950, when there was a very
large forest fire here in Canada, and clouds of smoke made it
all the way over to Europe. Through late September and early
October, people all over Europe observed blue moons and
blue suns. So it is quite a rare event and, from what reading I

have done on the subject, perhaps once every eighty or ninety years something like that may occur.

CALL ME MELLOW YELLOW
▲

Why is the sun yellow?

Dr. Peter Stetson, Senior Research Officer with
the National Research Council in Victoria:

To put the answer simply, the Sun is yellow because it is yellow-hot. We've all heard, of course, about things that are red-hot and things that are white-hot. Yellow-hot is between red-hot and white-hot. The colour is a direct reflection of the temperature. Whenever a solid object, a liquid, or a dense gas becomes sufficiently hot, the radiation it gives off will be visible, and the higher the temperature, the shorter the wavelength of the light.

The basic physics behind this is that as you heat up an object, the atoms in it move around faster and faster. Atoms and electrons bump into each other, and the speed of the collisions is determined by the temperature. When atoms and electrons collide, they give off light, and the wavelength of the light is determined by the strength of the collision, and therefore is also a function of temperature. The bulk of the light given off by these collisions is in wavelengths we can't see, infrared wavelengths, until the temperature rises to about three thousand degrees Kelvin, or above absolute zero. At that point, the bulk of the radiation reaches the wavelength of red light. As the temperature rises further, the light being emitted goes across the spectrum, through orange, yellow, green, blue, and violet.

So let's get to the sun. The temperature of the sun's surface is about six thousand degrees Kelvin, though it is hotter and colder in spots. At six thousand degrees, a typical collision involves enough energy to emit yellow light. Collisions that are more energetic or less energetic than average, of course, produce light in other wavelengths up and down the spectrum, but the dominant colour is yellow, and so that is what we see.

Interestingly, this changes just a little when we get out of the Earth's atmosphere. From space, the sun appears a little whiter than it does on Earth. Our atmosphere contains lots of dust particles and large molecules that tend to scatter the shorter wavelengths of light, which just bounce off them. Longer wavelengths are more able to pass through this natural filter. That is why we tend to see red sunsets. When the sun is closer to the horizon, sunlight is passing through a lot more air before it reaches our eyes, so even more of the shorter, bluer wavelengths are being filtered out, leaving the longer, redder wavelengths. That is also why you can look at the sun at sunset without being blinded quite so quickly – a lot of the light is being filtered out and so the sun seems much less bright.

Astronomers have discovered stars glowing in all sorts of colours. Some are as red as the coals in your fireplace and some are a beautiful aquamarine blue. It is simply a matter of what temperature they are.

LUNAR LONGEVITY

▲

How old is the moon and why can you see it in the daytime?

Terry Dickinson, Astronomer, Broadcaster, Author,
and Editor of *SkyNews* magazine.

The moon is 4.6 billion years old, the same age as the Earth
and all the other major bodies in our solar system. The solar
system was formed from a flat, pizza-shaped rotating cloud of
dust and gas. The sun formed at the centre of the cloud and the
planets formed in a flat plane around it. There was a lot of
debris and material left over during this formation, and the
moon was formed by an object about the size of the planet
Mars smashing into the Earth, grazing it. That collision
splashed material from both the Mars-like body and the Earth
into orbit around the Earth, and that material collected into
the moon. This probably happened within the first ninety
million years of the solar system's existence.

Much of this has been determined from the lunar rocks
brought back from the moon by the Apollo astronauts. Those
rocks have been dated, and this has given us a very good
handle on how old the moon and the Earth are. Interestingly,
we haven't been able to date the Earth's age using rocks on
Earth because the oldest rocks here are only about 3.9 billion
years old; anything older than that has been melted by the
much more active surface of the Earth. Earthquakes, volca-
noes, and the movement of the continents have destroyed all
the very ancient rocks on Earth, while the moon, which isn't

geologically active, has been essentially as we see it today for billions of years and the very oldest rocks are still around. So we learned the age of the Earth by examining the rocks from the moon.

Since the Apollo moon flights, we have had a good idea of the composition of the moon. Overall it is like the mantle of the Earth, closer to the surface, so that material in the moon rocks seems to have come from the surface of the Earth or another body like the Earth. The only way one can reasonably get that sort of material is by scraping off a chunk of the surface of the Earth or a similar body. Therefore, the collision scenario is the one that seems to explain best why we have a moon like that.

The listener's second question – Why can we see the moon in daylight? – is directly related to daytime sky conditions. On many days, the moon is, in fact, in the daytime sky, but we can't see it because of clouds, overall sky brightness, or glare from the sun. It takes special conditions to allow us to see clearly our companion world when the sun is in the sky. If you remember the last time you saw the moon by day, it was likely a clear, dry day with a deep blue sky, probably in late afternoon. Under such conditions, a minimum amount of light is being scattered in the atmosphere to obscure the relatively weak light of the moon. In other words, the silvery light of the moon has more contrast with the sky and is easier to see. A washed-out, milky, or hazy day sky that comes from increased high humidity or pollution tends to hide the moon.

A LIGHTWEIGHT QUESTION

▲

Are there any places in the universe where there is really zero, or null, gravity?

Chris Burns, Visiting Assistant Professor at Swarthmore
College in Pennsylvania, and former Ph.D. candidate in
the Department of Astronomy at the University of Toronto:

There are two answers to this question, a yes and a no, and
which one you choose depends on what you mean. If the
question is whether there is a place in space where there is just
absolutely no gravity, the short answer to that is no. The
reason is that, unlike a force such as electric force, where you
have a positive and a negative source so they can cancel each
other out, gravity only has mass, and we don't, as far as we
know, have anything that is anti-mass. So as long as you have
something in the universe, gravity is there, though its effects
can be very weak if you are far away from any mass.

If the question, on the other hand, is whether there are
points in the universe where it seems there is no gravity, that
is, you are not being accelerated in any direction, then the
answer is yes. In these places the pull of gravity from one
direction would be countered by an equal pull from an opposite direction, and the net effect on you would be zero. It's
not that there is no gravity, it is just that gravity is perfectly
balanced.

Let's take the example of the sun and the Earth. On the
surface of the Earth, of course, you feel the pull of the Earth far

more than that of the sun. As you move away from the Earth, the gravitational pull of the Earth will get weaker, and as you get closer to the sun, its pull will be stronger. It is easy to imagine that there's a certain point between the two where they're exactly equal. If you move just a little away from that point, you will be in a place where the balance is gone, and you will start to fall towards the one that is stronger. At the point of balance, you will feel no gravitation pull. These are actually called Lagrange points, after the eighteenth-century mathematician who first worked out where they should be. Rocket scientists have discovered that they are useful places to park satellites, because they can place a satellite there and leave it and it won't fall either towards the sun or the Earth. It is essentially a point in orbit around the sun that's also fixed relative to the Earth. One of the ideas they have for the next generation of space telescope is to place it at a Lagrange point, and it will stay there.

8

Hail and Hurricanes:

ENVIRONMENTAL SCIENCE AND WEATHER

HOT ICE

▲

How do you get hail on a hot summer's day?

Dr. George Isaac, Senior Scientist with the
Meteorological Service of Canada in Toronto:

On a summer day, when the air is hot, it usually contains a
lot of moisture. When it is hot, the air will rise and cool, and
water vapour begins to condense into water droplets. When
the water vapour condenses, latent heat is released, which
pushes the air up faster, carrying the droplets with it. These
droplets continue to get pushed up, to an area where the air is
colder than zero degrees Celsius, between two and four kilometres high. At this height, most of the droplets stay liquid, in
a form known as super-cooled water.

But some of the water droplets come into contact with
suspended particles floating around and freeze. These ice

particles are the nuclei of hailstones and they grow rapidly. As they travel through the cloud, more water droplets collect on the surface, making the hailstones larger and larger. But as water freezes onto the surface of the hailstones, more latent heat is released, increasing the updraft velocity. Sometimes a hailstone will start to fall when the updraft weakens, and then rise again when it falls into a stronger updraft. This up-and-down motion of the hailstones continues until they are too heavy to stay up in the cloud, and the hailstones fall to Earth.

It takes a pretty severe storm to make hailstones. You need one that goes well up into the atmosphere, on the order of ten to fifteen kilometres tall. It has to have very strong updrafts to keep pushing the water droplets up to the point where they freeze into stones, and then to keep supporting the stones as they grow.

We only see hailstorms during the summer. During the winter, there is not enough moisture and heat in the air to generate the latent-heat release you need to make hailstones. You also need a thunderstorm, something we rarely see in the wintertime, and even if you do, they are not the really deep thunderstorms you'd get in the summertime.

WRINGING OUT THE AIR

▲

How dry is it in Saskatchewan when the temperature reaches –40 degrees Celsius. I've heard we compare quite closely to the Sahara Desert. Is this true?

David Phillips, Senior Climatologist at Environment
Canada:

All Canadians know, and can clearly see any time we just
move across the carpet in the winter, that in winter it is stati-
cally dry in our homes and buildings. We actually add water
to our atmosphere with humidifiers to take away this dryness.
That is because, as the temperature falls, the amount of water
vapour in the air also drops.

Every parcel of air has water vapour; it is always present
in the atmosphere. Even in a bone-dry desert there is water
vapour. And the temperature tells you how much water vapour
you're going to have.

If you took a normal-sized room that was at saturation (in
other words, the air couldn't hold any more water without it
precipitating out), and filled it with air at 30 degrees Celsius,
you could wring about two litres of water out of the air in the
room. But if you took the same room and went down to
–20 degrees, which is a typical cold day in Winnipeg or
Saskatoon, you'd end up with about sixty-two millilitres of
water, or one-thirty-second of the water you found in the
warm room.

So you can see that cold air just can't hold as much water
as warm air, which is why it feels drier. It really *is* drier,
even though it is saturated with water at that temperature and
pressure.

If you want to compare the air in Saskatchewan in winter
to the air in the Sahara, you need to look at their different temperatures. The air over the Sahara is much warmer, so it can
support more water. I've read that if you wrung out the air in

both places, you'd get more water from the Sahara than you would from Saskatoon in January.

CYCLING THROUGH THE CYCLONE CIRCLE
▲

Most weather tracks from west to east, but hurricanes appear to start forming off the coast of Africa and track west, building power until they end up on the coast of North America. What is it that pulls these storms from east to west?

Guy Roussel, Meteorologist at the Canadian Hurricane
Centre of Environment Canada in Dartmouth, Nova Scotia:

The tropical cyclones that affect eastern North America originate in three areas: the Gulf of Mexico in the early season, the eastern Atlantic Ocean in mid-season, and the Caribbean Sea in the late season. "Tropical cyclone" is the generic term for the class of tropical weather systems known as tropical depressions, tropical storms, and hurricanes. Not all tropical cyclones will make landfall; sometimes they will dissipate over the ocean in their early stage of development. After moving up to the north of the horse-latitude region (the horse latitudes are at about 30 degrees north and south of the Equator), tropical cyclones begin interacting with frontal troughs, sometimes developing into hybrid storms extending over much larger areas. Tropical cyclones weaken rapidly when they move over land or over colder water.

Tropical cyclones tend to travel common paths until they dissipate. Their movement is characterized by their track, which is affected by several factors. First, they follow the predominant wind within a region. For example, late-season hurricanes over the Atlantic often begin in the tropical region near Africa between 10 and 20 degrees north, then drift west on the easterly winds called trade winds. Then the hurricanes veer northeast at about 30 to 35 degrees north in the region of the horse latitudes, where they meet the prevailing westerly winds blowing eastward across North America. These wind currents driving those storms are constantly changing, so the tracks are difficult to predict. The track is also affected by the Coriolis force (which makes winds in the northern hemisphere rotate counter-clockwise around a centre of low pressure), low-pressure troughs, and subtropical ridges. However, meteorologists, with the help of advanced computer models and sophisticated forecasting tools, succeed in predicting the intensity and tracks of those storms in a very reliable way.

RAINING HAMBURGER BUNS
▲

Why doesn't rain fall down in big lumps? Is it friction with the air, or surface tension of the water closing it into droplets?

Chris Doyle, Meteorologist with Environment Canada in Vancouver:

There are two theories about why rain doesn't fall in big lumps. One involves surface tension, and the other frictional forces. Water likes to stick together, and in the absence of any other forces it is going to stick together in a little ball. That is due to surface tension. But rain is a complex phenomenon. As well as surface tension, there are lots of other forces at work. The story of rain starts in the clouds, at the microscopic level. Initially, there are little particles in the air that are not water. Usually they are dust or pollution, and are called condensation nuclei. These nuclei hang about in the clouds and tend to attract water molecules. The water droplets latch on to the little pieces of dirt, and when the amount of water around the nucleus gets large enough, it forms a visible droplet. At that point the droplet is round.

The small water droplet is suspended by the updrafts in a cloud and it continues to attract water molecules. So it grows bigger and bigger for a while, but eventually the droplet gets too heavy to stay up there. Gravity starts to exert itself and overcomes the uplift from the cloud.

Now the droplets start to fall as rain. As each one starts to fall, it is a very small drop, in the order of less than a millimetre in diameter. Gradually it gets larger by colliding with other droplets on the way down. As it gets bigger it accelerates, and the aerodynamic forces start to deform the droplet from its round shape.

People might think the drop turns into the teardrop shape at this point, but that is not the case. It turns into something that looks more like an upside-down hamburger bun. As the drop approaches the ground, the front will begin to deform. It'll be swept back by the wind, making a round bottom, but there is a drag created on the sides of the drop. This drag creates little eddies behind it, which would actually tear off the little teardrop part we imagine, leaving the back of the drop flat.

The same aerodynamic forces that control the shape of the drops also limit their size. There is a whole field of study called cloud physics, and the scientists in that field have come out with numbers for these kinds of things. For a raindrop the size of three millimetres, the aerodynamic forces and the surface tension are just about in balance, but by about six millimetres, the aerodynamic forces exceed the surface tension and the drop fragments. It will break apart into smaller droplets and the process will start all over again, with the drops getting larger until they break apart again.

So we'll never see the kind of stream that comes out of a tap in the sink. But if you give them enough room, even these streams will eventually break up. If you think of a waterfall, the water comes over the edge of the cliff in a solid body, as a stream. But as it goes into the air, it accelerates down and the aerodynamic forces break it up. In the highest waterfalls, the water turns from a solid stream at the top to a fall of mist at the bottom.

So next time you're out in a real downpour, remember, it is not raining cats and dogs, but hamburger buns!

SNIFFING THE SEDIMENT

▲

What do I smell when it begins to rain? Is it ozone?

Dr. Phil Hultin, Associate Professor of Chemistry
at the University of Manitoba:

The answer to this question has to be dealt with in an indirect
way, because you smell many different things when it starts to
rain. One suggestion is that you smell ozone, but that is not
likely when a rainstorm just begins. People smell ozone
during a thunderstorm because ozone is formed by the electrical discharge we call lightning. High-voltage electricity
passing through the air forms ozone from oxygen. You might
get a whiff of ozone if you are downwind of an oncoming
thunderstorm. However, at the beginning of a gentle rainfall,
there is probably not a significant amount of ozone around.

I should probably note at this point that when we smell
something, what is happening is that relatively small molecules present in the air are binding to large protein molecules
that are part of the cell surfaces of our nasal passages. These
protein molecules are called receptors. Each receptor can bind
a limited class of small molecules that share common shapes.
When this happens, it triggers a chain of biochemical events
that sends a signal to our brains and we perceive a smell.

At the beginning of a rainstorm, there is a large amount of
material suspended in the air – a lot more than most people
imagine. The material is microscopic, and it is in the air

because of the action of the wind, which stirs this material up. It is composed of many different things, including pollen, spores, waxes, and resins from plants and leaves. There are also whole microbes and fungi, as well as humus from soil and debris from plant breakdown. Some of these airborne materials are present as individual molecules, while the larger particles are surrounded by a cloud of molecules characteristic of the material. They are all the things we smell.

When rain starts to fall, it pulls these particles out of the air. They stick to the water droplets and fall to the ground. When the raindrops hit the ground, they strike the surface with a relatively large amount of energy. This scatters the water back up into the air. The impact will also cause material from the ground to be sent up into the air. So, we probably see an increase in plant-derived and soil-derived materials during the first few minutes of a rainstorm. It is that earthy, planty, mushroomy kind of smell that we associate with rainstorms, but that usually goes away after the first few minutes of rain.

The earthy smell is what you are most likely to smell if you are in the country. But in the city, the smell is usually more like pavement. The same thing is happening in both situations. The difference is the material the water is hitting. Pavement is made of various hydrocarbon compounds called asphaltenes, which are obtained as byproducts in oil refining, combined with assorted filler material. It usually also has a generous surface coating of motor oil, transmission fluid, tire-rubber residues, particulates from exhaust emissions, and other oily materials dropped by our cars and trucks. These oil-based materials are thrown up into the air by the impact of the raindrops, so we tend to perceive their smell – hot, oily, and industrial – when the rain begins.

After the first few minutes of a rainstorm, the smell changes. At that point, the air usually smells fresh and clean. That is because the raindrops, as they fall through the air, scavenge the suspended material and actually wash it out onto the ground. After a few minutes of rain, the bulk of this material has been washed down to the ground and is out of circulation as far as our noses are concerned. What we are noticing then is the absence of odours that we have become accustomed to in our normal environment. They will be gone until the wind has a chance to blow them up into the air once again.

QUICKSILVER QUERY

▲

What happens to a mercury thermometer when the temperature goes below the numbers listed on it?

Peter Bowman, Instrumentation Specialist with the
Meteorological Service at Environment Canada in Toronto:

A mercury thermometer works because when you cool mercury (and most other liquids) its volume decreases. As the temperature drops, the mercury in a thermometer contracts, and the height of the column of mercury in the numbered stem of the thermometer drops. You may have noticed that most mercury thermometers only register down to about −37 degrees Celsius at the coldest. That is because mercury freezes into a solid at approximately −38.8 degrees C (−38.8344 degrees, if you want to know exactly). As a solid, it is not a hard metal like steel, but a soft one like tin or lead that can be cut easily with a knife. So when the temperature drops below −38.8 degrees Celsius, the mercury in a thermometer freezes, rendering it inoperable.

Of course, if the thermometer were water-filled, and the water froze, it would likely break because, unlike most other liquids, water expands when it freezes.

To read temperatures below −37 degrees Celsius, we frequently use alcohol or spirit-filled thermometers, because the freezing point of the specific liquid used is very low, in some cases as low as −115 degrees, much colder than we have ever recorded at the surface of the Earth.

The coldest temperature ever measured in Canada was −63 degrees, at Snag in the Yukon Territory in the 1940s. At that time, the spirit-filled thermometers in use did not register that low, since temperatures that cold had never been observed in Canada. In that instance, the observing staff at Snag scratched a mark on the glass of the thermometer stem where the top of the alcohol column had been observed, then returned it to headquarters for calibration.

Today, when we need to measure very cold temperatures accurately, we may use thermometers made from platinum, quartz, or some other kind of material whose electrical resistance is sensitive to changes in temperature. Using the platinum thermometers in my laboratory we are able to measure temperatures as low as −190 degrees Celsius. Of course, we still use spirit thermometers at many of our weather-observing stations, and likely will for some time to come.

BEWARE THE BOUNCING BALL

▲

A long time ago on a hill in Iowa, while on my way to Winnipeg, I was caught in a barrage of fireballs falling from the sky. They were about the size of a basketball. The first one seemed to fall from the low-hanging clouds, bounced two or three times, and exploded. When it blew up, it killed the engine in my car and I had to sit there for the next while, as these things bounced and blew up. I imagine that what I saw was what I later learned is called ball lightning. So my question is: what is ball lightning? How does it form and why does it act like it does?

Dr. Joe Kos, Professor of Physics at the University of Regina:

Science can only partly answer this question. In fact, nobody really understands what ball lightning is exactly. It is one of the most interesting and mysterious manifestations in nature, but not one that science has been able to examine.

The best guess is that ball lightning is a ball of plasma. Plasma is a gas that has so much energy that the electrons have been torn away from the atoms. Plasma glows because, as the electrons fall back, light is emitted. Neon lights work on this principle. The tubes are filled with gas, and electricity provides the energy to tear the electrons away from the atoms of gas. When they drop back, they release the energy we see in the typical colourful neon glow.

So ball lightning could be a glowing ball of plasma. The balls are often very bright and from their colour, their temperature has been estimated to be around 12,000 degrees Kelvin. The problem is that, at this temperature, the energy in a ball of plasma should be radiated within about a hundredth of a second, and the ball should collapse. But many descriptions of these balls have them persisting for several minutes.

Another mystery around ball lightning is that the balls often disappear with an explosion, and the energy associated with the explosion can be very large – much larger than we can explain in terms of the chemical energy of a ball the size of a basketball. We don't know where that energy comes from, though people have theorized that it could be contained within the ball itself, or pumped in from an external electromagnetic field. Part of the problem, and the mystery, is that ball lightning is very rare, and doesn't give scientists much of an opportunity to study it. The chances of you or me seeing it are very slim.

In fact, you probably don't want to see it, as it can be quite dangerous. People and animals have been killed by coming in contact with ball lightning. The listener who asked the question stayed in his car, so he was probably safe enough. There have been reports where ball lightning has entered homes, coming in through open doors and windows. In some cases it has caused considerable damage, but in others it has just disappeared quietly. It really is very mysterious stuff.

PRECIPITATION ACCELERATION

▲

How fast do raindrops fall?

Dr. Phil Austin, Associate Professor in the Atmospheric
Science Programme at the University of British Columbia:

It depends on how big they are. In a light drizzle, something
we're familiar with here in Vancouver, the smallest drops that
you can see are about the width of a human hair, and they fall
at about a kilometre per hour – not very fast at all. But in a
tropical downpour, or a big thunderstorm with sheet rain, the
drops can get as large as six millimetres, and they can fall at
the respectable speed of about thirty kilometres an hour. You
don't see drops larger than six millimetres because the stress
of air resistance breaks them apart.

The difference between one and thirty kilometres per hour
is pretty large. The explanation for that difference has to do with
the balance between the two forces acting on the raindrop:
gravity dragging it down to the ground, and the resistance of
the air that the drop is falling through. In the bigger drops,
gravity is a little more dominant because they have relatively
less surface area compared to their mass. The drizzle seems
sometimes to be hardly falling at all. In fact, drops smaller than
drizzle don't fall – they stay suspended in clouds, which are just
large accumulations of very fine droplets.

Very rarely, you will see drops falling faster than thirty kilometres per hour, but that's an unusual circumstance in which

the air is falling, as well as the rain. In those circumstances, as rain falls or as clouds mix with dry air outside the cloud, you'll get evaporative cooling of the rain. That evaporation can cool the air around the rain cloud, making it heavier than the dry air around it, and the air actually falls in a downdraft. The speed of a downdraft can reach eighty kilometres per hour, and can be very dangerous. These are the microbursts that you hear about, which can cause planes to crash. That happens if, on approach to landing, a plane passes through a precipitation shaft that has been cooled by evaporation. The plane can, all of a sudden, be pressed to the ground by a strong downward wind. In the past ten years, four or five very serious plane crashes have been caused by this phenomenon.

AN AIRY SILENCE
▲

What is the calm before a storm?

Dr. Phil Austin, Associate Professor in the Atmospheric
Science Programme at the University of British Columbia:

Apart from being a useful metaphor, the calm before the storm is a real meteorological event that is associated specifically with certain kinds of quite powerful storms. The kind of storm we are talking about would be the big towering thunder-storms that you typically see out on the high plains of the west, which cause heavy hail, rain, high winds, and possibly even tornadoes. These are really amazing objects. You can see them coming towards you at a distance on the prairies and they look like big black towers of cloud. The clouds themselves can be up to twelve kilometres tall and they have updrafts associated with them of fifty kilometres an hour. The storm basically sucks in air from the ground and pumps it upwards into its centre to feed itself.

A storm that is this big is able to organize itself so it has a vacuum intake region near its base and it is able to take in very warm, moist air that it is going to need to form the clouds with. Then it has got an exhaust region near its top, which goes up into the stratosphere. Pretty soon it is this big, flowing, anvil-shaped object.

Just as one of these storms approaches, you can feel the air go still and quiet and it is really very ominous. What is happening here is a cancellation of two winds. As you look at the storm coming towards you from a distance, you will probably feel a breeze in your face. That is the wind from part of the large weather system that is blowing the storm towards you. But as the storm gets closer and closer, you begin to become part of the storm's environment. The storm has to suck air in and that actually creates a wind in the opposite direction. If you are right at the base of the storm, you can be at a point at which the wind blowing the storm towards you is exactly cancelled by the movement of the air that the storm is sucking in to feed itself. This, together with the fact that the huge cirrus cloud is now right on top of you, blocking out the light, and the air is oppressive and humid, means that things can go very still. Birds will head for their nests, and you can get a real sense of foreboding.

This only happens if you are very close to the heart of the storm. So if you experience this, then you are really in the action centre of the storm. You can expect some pretty heavy rain, lots of thunder and lightning, and possibly hail.

These very powerful storms are pretty rare and they require special conditions. What you really need is very humid air, probably flowing north from the Gulf of Mexico. There is a lot of energy held in that humid air. Usually, you have a layer of much colder air overlying it that traps the humidity. When that layer of cold air breaks through, the hot air is much more buoyant and it flows upward with terrific energy and dumps all its water. That creates the big winds and the rain and hail with the kind of energy that can then get turned into tornadoes.

CLIMATE CATASTROPHE

▲

I've read that if global warming melts the Arctic ice cap, the added fresh water that would flow into the Atlantic would stop the Gulf Stream, driving Europe into a mini–Ice Age. Is this possible?

Dr. Andrew Weaver, Professor in the School of Earth and Ocean Sciences at the University of Victoria:

The scenario you mention appeared on the public's radar a couple of years ago, but it wasn't actually something that scientists working in the field had suggested was possible. It was a speculation by people who were knowledgeable, but not specialists, and the media found it interesting, so it got a lot of attention. It's also a scenario we get a lot of questions about from our students, who've all heard of it.

The idea isn't completely wrong. It's certainly true that we think global warming and melting ice caps can reduce the strength of the ocean currents that bring warm water from the equatorial oceans to the north, the North Atlantic Current and the Gulf Stream. It's even possible for these currents to be substantially reduced, but the implications for Europe would not be an Ice Age.

Normally, these currents bring warm water from the tropical ocean northward, where it releases its heat. The way warming might stop the current is by creating a cap on the ocean that would prevent sinking at high latitudes – a process that drives these currents. We're pretty sure this happened about eleven thousand years ago, in an event called the Younger Dryas. This was close to the end of the last Ice Age, and there were big ice sheets on land, both in Europe and Canada. Since it was the end of the Ice Age, temperatures were warming, and these ice sheets began melting. The result was that they dumped a tremendous amount of relatively cold freshwater onto the top of the ocean. Freshwater is lighter than seawater, so it floated on top of the ocean, and basically blocked the sinking of water at high latitudes. The warm currents were then stalled. The result of this was that Europe went back into a mini–Ice Age, just as it was coming out from the depths of the last Ice Age.

You might wonder, if this happened once, why it couldn't happen today. The important difference is that we're not coming out of an Ice Age at the moment. We're in a warm period that is actually getting warmer. The amount of fresh-water to melt is smaller, and even if it were to slow the warm southerly currents, the effects would likely be much less severe.

There have been several computer simulations, run by different groups, that all point to the fact that Europe will not go into an Ice Age. In fact, it will probably not even get cooler. Some of the simulations show the Gulf Stream being reduced just a little, some of them have it not being reduced at all, and some of them have it being dramatically reduced. In all cases, you still get warming in Europe. It's just warming a little slower than the rest of the world.

ACKNOWLEDGEMENTS

▲

Thanks to:

- *Quirks & Quarks* Senior Producer Jim Handman, whose dogged determination and alluring alliteration helped to guide this project from conception to completion, from punctuation to publication;

- *Q&Q* producer Jim Lebans, whose searing satire and wicked wit have lifted the writing on many a *Question Show* from mundane to marvellous; and whose literary skills helped to make the answers in this book both comprehensive and comprehensible;

- *Q&Q* producer Patric Senson, (the only one of us who is actually a trained scientist), whose scientific savvy and humane humour turned many answers in this book from gibberish jargon to proper prose;

- the many *Q&Q* producers, researchers, and interns, who, over the past ten years, have persistently pursued experts from across the country to answer your quizzical queries;

- all the scientists, researchers and professors who generously gave their time to answer our perplexing puzzles, and who graciously gave permission to include their answers in this book;

- and last, but certainly not least, to all of you who sent us your questions, and without whom there would not be a *Question Book*.

How to Reach Us

▲

If you have a question you'd like answered or have a comment on the program or this book, let us know. You can reach us at:

E-mail:	quirks@toronto.cbc.ca
Phone:	(416) 205-6131
Fax:	(416) 205-2372
Mail:	Quirks and Quarks
	P.O. Box 500, Station A
	Toronto, ON
	M5W 1E6